Lasers and Their Prospects

N. Sobolev

Translated from the Russian by V. PURTO

University Press of the Pacific
Honolulu, Hawaii

Delta College Library
November 2004

Lasers and Their Prospects

by N. Sobolev

Translated from the Russian
by V. Purto

ISBN 0-89875-043-1

Copyright © 2000 by University Press of the Pacific

Reprinted from the 1974 Russian edition

University Press of the Pacific
Honolulu, Hawaii
http://www.UniversityPressofthePacific.com

All rights reserved, including the right to reproduce this book, or portions threof, in any form.

TK 7871.3 .S613 2000

Sobolev, Nikolaæi
  Aleksandrovich, fl. 1968-

Lasers and their prospects

Contents

Introduction	7
Chapter 1. The Nature of Light	13
Particles or Waves?	13
Photons—Quanta of Light	18
Chapter 2. Atom as Quantum System	23
Quantum Concepts on the Atomic Structure	23
Structure of the Atom. Quantum Numbers	25
Radiation and Absorption	31
Distribution of Particles Among Energy Levels	36
Active Systems	38
Chapter 3. Generators of Light	43
Ruby Laser	43
Properties of Laser Beam	51
Active Materials	56
Methods and Sources of Excitation	63
Resonant Systems	70
Continuous-Wave Lasers	74
Glass Lasers	77
Giant Pulses	80
Gas Lasers	85
Methods of Concentrating Gas Laser Radiation at One Frequency	101
Liquid Lasers	104
Semiconductor Lasers	106
Chapter 4. Application of Lasers	117
Lasers in Communications	117
Light Beam Modulation Methods	120
Beam Waveguides	132
Lasers in Computers	138
Application of Lasers in Metrology	143

Lasers in Chemistry 145
Lasers in Photography 147
Lasers for Treating of Materials 155
Laser Gyroscopes 159
Lasers in Detection and Ranging 163
Laser Range Finders 167
Laser Tracking of Satellites 170
Lasers in Space Equipment 177
Communication with Spacecraft During Atmospheric Re-Entry 182
Detection and Communications Under the Sea . 185
Other Military Applications of Lasers 188
Lasers in Medicine and Biology 194

Chapter 5. Lasers and Science 200
Testing Einstein's Theory of Relativity . . . 200
Measuring the Drift of Continents by Means of Lasers 204
Lasers for Geodetic Studies and Atmospheric Sounding 208
Measuring of Speeds 210
Laser Space Communications 210

Chapter 6. The Prospects of Lasers 225
Pipeline out of a Laser Beam 225
Lasers and Communications with Extraterrestrial Civilizations 228
Spaceship of the Future 241
Looking Ahead 243

Introduction

Lasers are one of the biggest achievements made in the second half of the twentieth century.

Lasers are quantum generators working in the optical region of the spectrum, or, simply, generators of light.

The principle on which their work is based is the amplification of electromagnetic oscillations by means of forced or induced radiation of atoms and molecules. This kind of radiation was predicted by Albert Einstein as long ago as 1917 when he studied the equilibrium between the energy of atomic systems and their radiation. Therefore it would be, perhaps, true to say that the history of the creation of lasers begins just as early.

Yet, at that time nobody was aware of the potential value of this phenomenon for the future. Nobody knew then how induced radiation could be brought about and used.

In 1940, the Soviet scientist V. Fabrikant, analysing the gas discharge spectrum, pointed out that induced radiation effect could be employed for attaining amplification of light. In 1951, together with F. Butayeva and M. Vudynsky, V. Fabrikant carried out first experiments in this direction.

In 1952, scientists in three countries—N. Basov and A. Prokhorov in the USSR, Ch. Townes, J. Gordon and H. Zeiger in the USA, and J. Weber in Canada—simultaneously and independently suggested a new principle of generating and amplifying microwave frequency electromagnetic oscillations, based on the use of the induced radiation phenomenon. This allowed the creation of quantum generators for the centimetre and decimetre frequency bands, now known as masers; these devices displayed a very high stability of the frequency. The use of masers as amplifiers provided for a more than hundred-fold increase in the sensitivity of radio receiving equipment. First quantum generators employed two-level energy systems and spatial separating of molecules with different energy levels in a non-uniform electric field. In 1955, N. Basov and A. Prokhorov suggested that particles could be brought to the non-equilibrium state by using quantum systems with three energy levels, excited by an external electromagnetic field.

In 1958, N. Basov, B. Vul, J. Popov and A. Prokhorov in the USSR and Ch. Townes and A. Schawlow in the USA investigated the possibility of applying this method for developing optical generators.

Proceeding from the results of these investigations, in December 1960, T. Maiman of the USA constructed the first optical quantum generator which could successfully operate and in which synthetic ruby was employed as the active medium.

The first ruby quantum generator worked pulsed. Its radiation lay in the red region of the optical spectrum. Excitation was effected by means of a powerful light source.

With the appearance of the ruby optical quantum generator, the word "laser" has come into existence, which is an abbreviation based on the English description of the function of this device, Light Amplification by Stimulated Emission of Radiation.

A year later, in 1961, the American scientists A. Javan, W. Bennett and D. Herriott constructed a gas laser with a mixture of helium and neon as the active medium. The active medium in this laser was excited by the electromagnetic field of a high-frequency generator. This was a continuously working or continuous-wave (CW) laser.

The fact that induced radiation was obtained in a semiconductor diode in both the Soviet Union and the United States (1962) heralded the advent of a semiconductor laser. The possibility of employing semiconductors as the active medium in lasers was first pointed out by the Soviet scientists N. Basov, B. Vul and J. Popov as early as 1959. Much credit for the development of the semiconductor laser is also due to the American scientist R. Hall. The semiconductor laser is excited directly by electric current. This laser is capable of working both pulsed and CW.

At present more than a hundred of substances are used in lasers as the active media. Generation has been obtained with crystals, activated glasses, plastics, gases, liquids, semiconductors, plasma. As active media use can be made of organic compounds activated with ions of rare-earth elements. Success has been made in obtaining generation by employing common water vapours and even air. A new class of gas lasers, so-called ion lasers, have been created.

The operating wavelength range of the present-day optical quantum generators extends from 0.23 to 538 μ (from $1.3 \cdot 10^{15}$ to $5.57 \cdot 10^{11}$ Hz). The scale of the spectrum for electromagnetic radiation in the optical region is shown in Table 1.

Table 1

Scale of Spectrum of Electromagnetic Radiation in Optical Region

Ranges of optical region	Wavelength. λ, Å	Frequency f, T
Infra-red:		
long-wave range	$7.5 \cdot 10^6 - 2.5 \cdot 10^5$	$0.4 - 12.0$
medium-wave range	$2.5 \cdot 10^5 - 2.5 \cdot 10^4$	$12.0 - 120$
short-wave range	$2.5 \cdot 10^4 - 7.6 \cdot 10^3$	$120 - 400$
Visible:		
red	$7600 - 6200$	$400 - 485$
orange	$6200 - 5900$	$485 - 509$
yellow	$5900 - 5600$	$509 - 537$
green	$5600 - 5000$	$537 - 600$
blue	$5000 - 4800$	$600 - 625$
indigo	$4800 - 4500$	$625 - 668$
violet	$4500 - 4000$	$668 - 750$
Ultra-violet	$4000 - 50$	$750 - 6 \cdot 10^4$

1 Å = $10^{-4} \mu$ = 10^{-8} cm
1 T (one terahertz) = 10^{12} Hz

But what is the key advantage offered by these devices? This advantage resides in a number of remarkable properties of laser radiation. In contrast to light emitted by conventional sources, this radiation is *coherent with space and time, it is monochromatic, propagates in the form of a very narrow beam and is characterised by an extremely high concentration of energy* which only recently seemed fantastic. All this makes the laser beam a finest instrument in the hands of scientists for the investigation of various substances, for the study of the characteristic properties of atomic and molecular structures, for a better understanding of their interaction, for the determination of the biological structure of living cells.

Laser beam can be used to transmit signals for communication both on the Earth and in space practically over any distance.

It is quite obvious that in the nearest future communications with space rockets and spaceships will be effected with the help of lasers. The use of lasers in terrestrial communications will bring about a real revolution in the communication engineering. The amount of information that can be transmitted via communication lines will grow immensely. It suffices to say in this connection that theoretically one laser beam offers a channel for about one thousand million telephone conversations to be carried out simultaneously.

Lasers make possible radar observations of celestial bodies. In November 1963 a first experiment was conducted in the Soviet Union to explore the Moon with the help of a laser. The radiation of the laser installed in the focus of the 2.6-m telescope of the Crimean Astrophysical Observatory was sent in the form of powerful pulses towards the Moon. The beam reflected from the Moon returned to the Earth and was detected by very sensitive receiving devices. The intensity of the reflected beam was 10^{19} times weaker than that of the original beam. Yet it proved to be sufficient for determining the configuration of the natural Earth's satellite and the dimensions of separate areas of the lunar surface with an accuracy more than one hundred times exceeding that attainable with any other method, however perfect it may be. The depth of the lunar craters was determined in this way.

Lasers hold the greatest promise for diverse technical uses. They can find application in ranging and navigation, in medicine and biology, in chemistry and geophysics. Lasers will also be widely employed in industrial treatment of materials. Apparatus have already been created for performing various technological operations which could hardly, if at all, be performed before. For example, rough finishing of a hole in a diamond die usually takes more than two hours; with the use of a laser this operation can be completed in less than one tenth of a second.

Our time is that of space travels. But before starting space journeys to other distant worlds men should make certain about the existence of extra-terrestrial civilizations. If such civilizations really exist somewhere in the Universe, then, quite probably, lasers will help us to get into communication with them.

Taking into account that lasers were created only a little more than ten years ago, to-day it is impossible to predict all those fields in which they can and actually will be employed. Yet there can be no doubt that these wonderful devices have a great future.

The possibility of using optical quantum generators in various branches of science and technology is under active investigation now. The creation of optical quantum generators owes much to Soviet scientists. In 1959, N. Basov and A. Prokhorov were awarded the Lenin Prize for the research work in the field of quantum electronics. In recognition of the fundamental research in quantum radiophysics which led to the creation of the new type of generators and amplifiers—lasers and masers, the Swedish Academy of Sciences awarded N. Basov and A. Prokhorov of the USSR and Ch. Townes of the USA, by the Nobel Prize for physics of 1964.

CHAPTER 1

The Nature of Light

PARTICLES OR WAVES?

The world in which we live is full of light. Light is radiated by the Sun, by stars, by glowing electric lamps, by a burning match, and by dazzling flashes of lightning. Light enables us to see the beauty of the universe around. But what is light? What are the nature and structure of it? What processes in matter cause light radiation? These and many other problems have always been of interest to man, but it took centuries before he could riddle them. And small wonder, since an insight into the nature of light gives a clue to the understanding of matter.

The earliest investigations into the nature and behaviour of light were made by the ancient Greeks. According to one of the theories which they put forward, light emanated from the eyes of man and therefore he could see objects around him.

The well-known German astronomer Johann Kepler who lived in the seventeenth century considered that light was a kind of substance continually emitted by light-radiating bodies. He was of opinion that the propagation of light was instantaneous.

The first definite view-point on the nature of light

was formulated by Isaac Newton. He held that light is a flux of specific material particles (corpuscles) which are emitted by luminous bodies and propagate in a homogeneous medium in straight lines with a definite finite velocity. "By the rays of light I understand its least parts", Newton wrote, convinced in the correctness of his idea.

Newton was the first to observe dispersion of light with the help of a prism and to give an explanation of this phenomenon. He explained colour by the sizes of corpuscles which produced it. "Nothing more is requisite for producing all the variety of colours", Newton wrote, "than that the rays of light be bodies of different sizes, the least of which may make violet ... and the rest as they are bigger and bigger, may make ... blue, green, yellow and red." Depending on their sizes, corpuscles travelled with different velocities. Newton held that the velocity of light was also dependent on the medium in which it propagated: that it was greater in a denser medium and smaller in a medium of a lower density. He came to this conclusion when trying to explain the laws of refraction and reflection of light. As we shall see later, Newton proved to be right only in the first part of his assertion.

Thus in the end of the seventeenth century the corpuscular theory of light, otherwise called "Newton's theory of emission", came into being. It gained wide recognition and reigned supreme in science almost throughout the eighteenth century. Though generally accepted, the corpuscular theory could not explain many of the phenomena in the behaviour of light, such as diffraction, interference and polarization. Such weak points in the corpuscular theory made even Newton's contemporaries feel somewhat dissatisfied with it.

In the same seventeenth century, only some years after Newton had formulated his theory of emission, the Dutch physicist Christian Huygens proposed an undulatory theory of light, which was to become the irreconcilable rival of the corpuscular theory. Huygens denied the existence of light corpuscles. According to his theory, light had a vibrational character and was a kind of elastic impulse propagating in a specific medium, the aether, which pervaded all space. Huygens' theory postulated the presence of the aether in water, air, glass, and even in vacuum. In Huygens' opinion, the propagation of light in the aether was analogous to the propagation of sound in air, the both phenomena being of undulatory character.

When explaining the refraction of light, Huygens, as well as Newton, proceeded from the difference in light propagation velocities in different media. But the conclusions drawn by Huygens were diametrically opposite to those made by Newton.

Huygens' theory, however, did not become widespread at that time. In the "age of Newton" not every scientist was bold enough to doubt the rightness of the views professed by the man of the highest scientific authority, all the more so that Newton himself discarded Huygens' theory.

In the late eighteenth and early nineteenth centuries some scientists could quite successfully account for a number of phenomena with the help of the wave theory of light. Thus, after the British scientist Thomas Young had thoroughly investigated the interference and diffraction of light, the French scientist Augustin Fresnel gave full theoretical interpretation of these phenomena, reasoning from the wave theory. Fresnel offered a consistent explanation of all the experimental data on the diffraction and interference known by that time. He also put forward the idea that light vibrations were transverse, which permitted the polarization phenomenon to be understood and explained.

By then the velocity of light was already known to be finite. The Danish astronomer Olaf Römer was the first to establish this fact in 1675. He also succeeded in fixing a definite value for the velocity of light by observing eclipses of Jupiter's satellites. This value proved to be 300 000 km/s. In 1849, the French physicist Armand Fizeau measured the velocity of light in air. One year later another French physicist Léon Foucalt determined the velocity of light in water. It was found that the velocity of light in water was approximately 1.33 times less than in air. For the first time the validity of Newton's and Huygens' hypotheses on the refraction of light at the boundary between two media could be practically tested. Foucalt's experiment on the velocity of light in water was crucial between the two theories: it did not confirm Newton's hypothesis and decided in favour of the wave theory.

Thus, in the end of the nineteenth century, the wave theory of light at last won recognition.

But the nature of light was still an enigma. As before, the corpuscular theory and the wave theory had their adherents. Quite a number of phenomena could be well explained from the standpoint of the wave theory, while the corpuscular theory could offer no explanation for them; other phenomena, on the contrary, could well be described with the help of the corpuscular theory but could not be described at all in terms of the wave theory.

In the second half of the nineteenth century the famous British theoretical physicist James Clerk Maxwell constructed his harmonious theory of the electromagnetic field, which was based on the discoveries made by Coulomb, Ampère and Faraday in electricity. According to this theory, the electromagnetic field was a specific all-pervading medium. Maxwell succeeded in describing all the laws governing electromagnetic phenomena by a comprehensive system of equations. On the strength of these equations he assumed that light should be regarded as a variety of electromagnetic vibrations. This was indeed a brilliant idea. Calculations showed that electromagnetic waves propagate with a velocity of 300 000 km/s, which is exactly the same as that of light. The connection between two fields of physics—light and electricity—was thus discovered.

So, light wave is a wave of the electromagnetic field. This wave is characterized by the electric field vector E and the magnetic field vector H, that are at right angles to each other, and by the velocity of propagation u.

The electromagnetic wave is transverse, which means that both field vectors are oscillating in time perpendicularly to the direction of their propagation, in the same phase (Fig. 1).

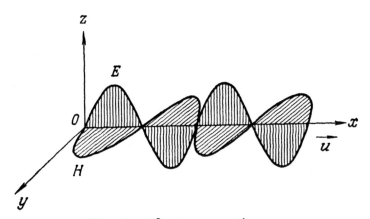

Fig. 1. Electromagnetic wave

Experiments carried out by the German physicist H. Hertz on the production of electromagnetic waves and studies of their properties confirmed the deep analogy between electromagnetic and light waves. No one could be any longer in doubt as to the wave nature of light.

The discovery of the electromagnetic nature of light disproved Huygens' hypothesis of the "elastic aether". Nevertheless, the conceptions of the aether as the medium for the propagation of electromagnetic waves were still entertained, only the aether was thought to be not "elastic", but "electromagnetic".

The untenability of the hypothesis of the aether as the only and quite specific medium in which electromagnetic waves (light) can propagate became evident only after the advent of Albert Einstein's theory of relativity in 1905.

PHOTONS — QUANTA OF LIGHT

The development of physical concepts on the nature of light during the period of two and a half centuries was complicated and even unexpected in many respects. Physicists kept on investigating and searching. New experiments were made, new facts became known, and new outlooks on the nature of light appeared.

For conducting experimental research of the laws of thermal radiation, the German physicist Gustav Kirchhoff suggested a model of a perfectly black body.

While studying the radiation characteristics of different bodies at different temperatures, the German scientists Josef Stefan and Ludwig Boltzmann discovered the law which states that the radiant emittance E of a perfectly black body is proportional to the fourth power of its absolute temperature T; $E=\sigma T^4$, where σ is a constant.

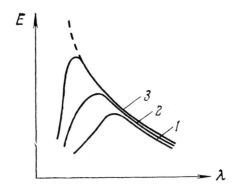

Fig. 2. Dependence of spectral concentration of radiant emittance on wavelength

Another law, known as Wien's displacement law, states that the wavelength in the radiation spectrum of a black body, corresponding to the maximum spectrum density of radiant emittance, λ_{max}, is inversely proportional to its absolute temperature: $\lambda_{max} = c/T$, where c is a constant.

Wavelength curves for the distribution of radiant energy at different temperature are shown in Fig. 2. Curve *1* corresponds to a relatively lower temperature, curve *2*, to a medium temperature, and curve *3*, to a higher temperature. The area defined by each of the curves is equal to the total energy radiated by the surface of a perfectly black body per unit of time.

As can be seen from this figure, for higher temperatures the radiation maximum shifts to the left, towards shorter wavelengths.

The British scientists J. Rayleigh and J. Jeans derived a formula trying to combine the above two laws. According to this formula, the radiant emittance of a hot body is directly proportional to its absolute temperature and inversely proportional to the square of the wavelength of light emitted by it.

This formula was in good agreement with the experimental data, but only in the long wave region of the visible spectrum. For waves of the ultra-violet region no such agreement could be observed. From the Rayleigh—Jeans formula it followed that the shorter the wavelength, the higher the radiant emittance, i.e. the intensity of thermal radiation, should be. With the transition to shorter waves the intensity should increase indefinitely. This, however, was in contradiction with experimental data.

In Fig. 2 the dashed curve corresponds to the combined Rayleigh—Jeans formula. It fits the experimental data obtained for the region of long waves but is drastically at variance with those obtained for the region of shortest waves. While the experimental curve drops to zero within this region, the curve based on the laws of classical physics tends to infinity. The situation which thus arose in the theory of radiation was termed by the physicists "ultra-violet catastrophe".

To explain the laws of thermal radiation, in 1900 the German physicist Max Planck made a quite daring assumption: he suggested that energy in the form of electromagnetic waves must be emitted not continually, but only in discrete amounts — quanta. This was in contradiction with the classical laws of physics, but using this idea M. Planck was able to derive an equation which established the dependence of intensity on frequency, which was in excellent agreement with the experiment. According to Planck, for the frequency ν, the energy of the quantum is $h\nu$, where h is a constant, quantum of action, called *Planck's constant*. Its value is $6.625 \cdot 10^{-27}$ erg-s. Planck's constant is a fundamental quantity which was to play the greatest role in the entire physics. The very idea put forward by Planck about the discontinuous, discrete character of radiation was revolutionary for the physics of that time. It ushered in the period of transition from the old tenets of classical physics to quantum mechanics. Later on it led to the creation of the photon theory of light.

The introduction of the idea of discreteness into the interpretation of phenomena which at that time seemed to be typically continuous suggested that electromagnetic waves must also partake of the nature of particles. The results of investigations into photoelectric effect, which brought the classical theory to a deadlock, could be explained, as was shown by A. Einstein, only with the help of the quantum theory, assuming that the energy of electromagnetic waves can be absorbed and emitted in quanta, each with the energy of $h\nu$. Quanta display the properties of particles. To emphasize this fact, light quanta were called *photons*. Thus, in the beginning of the twentieth century, physicists again turned to Newton's views on the corpuscular nature of light. The French physicist and astronomer D. Arago was right indeed in remarking that one should never neglect the foresight and conjectures of great men.

The main characteristic of light quanta—photons—is the amount of energy associated with them. Monochromatic luminous flux consists of quanta having the same energy. The frequency of radiation, according to the concepts of the quantum theory, is characterized just by the energy of the quanta E.

According to the wave theory, different kinds of vibrations differ from one another by the frequency of vibrations ν, this being the main parameter characterizing the wave process. The relationship established between these quantities is $E = h\nu$.

The intensity of the luminous flux is determined by the number of quanta traversing one square centimetre of a surface area per second. Since any luminous flux is made up of separate quanta, bodies are capable of emitting and absorbing light only in amounts that are integral multiples of the fundamental energy unit $h\nu$.

The mass of the light quantum manifests itself only when the latter is in motion; in the state of rest this mass is zero. It should be pointed out that the quantum properties of light show up the stronger, the greater its energy and mass are.

The picture created by scientists at the present stage of development of the theory of light reflects the dialectical unity of its contradictory corpuscular and wave properties. Yet, even to-day we cannot say that this picture is complete.

The light quanta, how and where do they originate?

CHAPTER 2

Atom as Quantum System

QUANTUM CONCEPTS ON THE ATOMIC STRUCTURE

The first model of the atom, fundamental for our present-day concepts of its structure, is due to the great British physicist Ernest Rutherford. Rutherford suggested that the atom consists of a positively charged nucleus in which almost the whole mass of the atom is concentrated and of negatively charged electrons revolving about the nucleus in certain orbits. This model was advanced by Rutherford as a result of his numerous experiments on the bombardment of targets made of different elements with thin bunches of helium nuclei, carried out in 1911.

At first sight Rutherford's model of the atom has much in common with the model of the solar system. For this very reason Rutherford's model of the atom was called planetary.

Rutherford's model, however, was not free from certain disadvantages. For instance, it could furnish no explanation of the exceptional stability of the atom. Reasoning from the laws of classical physics, the revolution of electrons about the nucleus cannot be stable, since, as any accelerated motion of charged particles, it must be accompanied by electromagnetic radiation.

An electron moving in a circular orbit, even with a constant speed, possesses an acceleration, according to the laws of classical physics. For an electromagnetic field to be set up a certain amount of energy is to be expended. Therefore the energy of the electron must gradually diminish and the speed of its motion gradually decrease. Eventually the electron should spiral down into the nucleus. It can be calculated that with an atom of hydrogen this process would be completed in about 10^{-8} s. Experimental, practical evidence, however, does not confirm such behaviour of electrons. On the contrary, atoms are very stable and can exist for as long as many a thousand million years.

In 1913, the Danish physicist Niels Bohr was successful to find the correct way out of this difficulty and explain the origin of line spectra of different elements, as well as the stability of the atom. Bohr showed that the laws of classical physics could not be applied to intra-atomic processes and that such processes should be interpreted in terms of the quantum theory.

He maintained that the electron in the atom is restricted to particular stable orbits (or shells) which are at different distances from the nucleus; the electron can never be found between such orbits. As long as the electron remains in steady, definite energy states, it does not radiate or absorb electromagnetic waves. The electron can pass from its one steady state to another only in a jump. Such transitions are accompanied by radiation or absorption of electromagnetic waves.

With the atom passing from one steady state of energy E_2 to another one E_1, the radiation or absorption of electromagnetic waves is always in integral quanta only, and the frequency of radiation (absorption) ν, multiple of Planck's constant h, is given

by the formula
$$E_2 - E_1 = h\nu$$

As can be seen from this formula, the frequency of radiation depends only on the difference between the energies the atom had in its respective energy states, whereas, according to classical physics, the frequency of radiation is in no way related to the amount of the energy radiated.

The Bohr theory could well describe the discrete spectrum of the one-electron atom of hydrogen, but could not account for the spectrum of an atom as simple as helium having two electrons. Nor could Bohr's theory explain the relationship between the intensities of different lines in the atomic spectrum. Nevertheless, the quantum theory of Niels Bohr offered a way to the understanding of complicated processes of atomic radiation and, following the theory of Albert Einstein to whom we owe our present-day views on the nature of light, for the first time answered the question how and where light quanta originated.

The quantum theory or, as it is more often called, *quantum mechanics*, was later developed into a still more rigorous logical system by such prominent scientists as Arnold Sommerfeld, Erwin Schrödinger, Werner K. Heisenberg, and others.

STRUCTURE OF THE ATOM. QUANTUM NUMBERS

Let us consider the structure of the atom taking as an example the simplest case of a hydrogen atom.

A hydrogen atom consists of a nucleus (proton) and an electron moving about the nucleus in definite orbits (shells) (Fig. 3). The electron and the proton are charged particles and their charges are equal in magnitude but opposite in sign. As a whole, the hydrogen atom is electrically neutral. The nucleus of hydro-

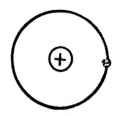

Fig. 3. Model of hydrogen atom

gen and its electron are mutually attracted by the electrostatic force and therefore the electron does not fly away from the nucleus.

To make the characteristic of the hydrogen atom more complete, it should be mentioned that the mass of its nucleus is 1 836 times that of the electron. The mass of the hydrogen atom is thus practically determined by the mass of its nucleus (proton). The mass of the atom is $1.67 \cdot 10^{-24}$ g. The diameter of a hydrogen atom is approximately equal to 10^{-8} cm. This value corresponds to one angström. The dimensions of an atom cannot be determined precisely: its boundaries are somewhat fuzzy.

The radius of the nucleus of a hydrogen atom is about one hundred thousandth that of the atom and equals $1.3 \cdot 10^{-13}$ cm. The density of the substance in the nucleus is extremely high: it comes to about $2 \cdot 10^{14}$ g/cm^3, that is, to about two hundred million tons per cubic centimetre. If we could make a pinhead from a substance with such a density, it would be heavier than a block of iron as big as a ten-storey building.

The orbits or shells an electron can occupy in an atom are designated by letters K, L, M, N, etc. Therefore an electron occupying the shell nearest to the nucleus is called K-electron. The shells can be numbered by ascribing respective numeric symbols 1, 2, 3, 4, etc. to them. These numbers are called *principal quantum numbers* and denoted by the symbol n.

The lowest orbit, nearest to the nucleus, for which $n=1$, is the most stable orbit for a hydrogen atom; and an electron in this orbit or shell is said to be in its *ground state.* The energy the electron possesses in this shell is characterized by a certain value E_1. For the electron to be transferred to another shell, more remote from the nucleus, a quite definite amount of energy should be imparted to the electron. In a particular case this energy may be a light quantum, photon. In another shell the energy of the electron will be E_2, equal to the energy the electron had in the previous shell plus that of the photon.

All the states of a hydrogen atom, when the electron occupies a shell other than the nearest to the nucleus, are called *excited states.* If we relate this concept to the principal quantum numbers, excited states are those with the principal quantum number greater than unity. The radii of a hydrogen atom in different excited states are proportional to the square of the principal quantum number.

The atom cannot reside in an excited state for a long period of time and tends to return to its normal, stable state with minimum energy. With the electron returning to its initial orbit, the atom emits the same amount of energy it had received, as a quantum of electromagnetic radiation $h\nu$. Similar transitions in the atom can take place between other orbits as well. As a result, there is a whole series of frequencies constituting the emission spectrum of an atom. Every atom has its own strictly definite spectrum of frequencies. The more complicated the structure of the atom, the more complicated its spectrum is.

Figure 4 is a diagram which shows the energy levels and the corresponding radiation frequencies of a hydrogen atom. Possible transitions are indicated by vertical arrows. Numbers at the arrows identify the radia-

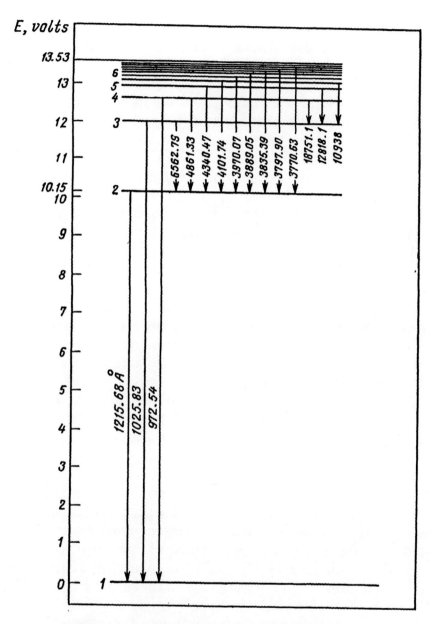

Fig. 4. Energy levels of hydrogen atom

tion wavelength corresponding to the energy of a given transition.

Not only the energy of an electron, but other characteristics of the atom are also quantised.

The electron possesses an angular momentum. As is known from mechanics, angular momentum of a particle is the product of its mass, velocity and the distance of this mass from the centre about which it rotates. According to quantum mechanics, the angular momentum of an electron is quantised as well, i.e. it can have not any arbitrary, but quite definite values, known as *orbital* (or asimuthal) *quantum numbers*. These numbers are denoted by the letter l. The maximum value of the orbital quantum number for a given shell is equal to the principal quantum number minus unity. For example, for the shell M $n=3$. The possible values of the orbital quantum numbers for this shell are $l=2$, $l=1$, $l=0$. The states corresponding to the values of $l=0, 1, 2, 3$, etc. are denoted by respective letters *s*, *p*, *d*, *f*, etc.

The principal quantum numbers characterize the value of the energy of the electron, which depends on the radial distance of the electron from the nucleus. Orbital quantum numbers express possible values of the angular momentum of the electron in the orbit.

Besides these, there are two more quantum numbers, so that the total number of possible states of an electron in the atom is still greater. It is known that an electron, while in orbit, creates electric current and, as a result, a magnetic field is set up. The magnitude of the magnetic field due to circular current is characterized by a magnetic moment. If an atom is placed into an external magnetic field, the direction of the magnetic moment of the orbital current may happen to be at a certain angle to this field. The smaller the angle of inclination, the greater the projection of the magnetic moment of the orbital current onto the direction (vector) of the external field will be.

The projection of the orbital moment onto the vector of the external magnetic field is also a quantised variable. For the orbital quantum number equal to l, the magnetic quantum number can take up all values from l to $-l$ differing from one another by unity. In the magnetic field the sub-level corresponding to the orbital number l consists of $2l+1$ states which are characterized by different magnetic quantum numbers. These numbers are denoted by the symbol m_l.

The electron has its own angular momentum termed the *spin*. The spin is also a quantised variable. It can be either parallel or anti-parallel to the orbital momentum. The spin quantum number is denoted by the symbol m_z.

We thus have to deal with a series of different quantum states of the atom. The picture of energy levels for atoms with a large number of electrons is extremely complicated.

With a system consisting of atoms, molecules and ions the picture of the distribution of energy levels differs substantially from that we have for an individual atom, since the electromagnetic fields of separate particles interact with one another. In solids the neighbouring atoms are found so close to each other that their outer shells are in contact and even overlap. The interaction of the electron shells results in a shift of the energy levels and in the formation of energy zones of a definite width. Gases in which the interaction between the particles is weak, have narrow energy bands. In solids, instead of narrow energy levels characteristic of individual atoms, broad energy zones are observed.

RADIATION AND ABSORPTION

Let us discuss quantum transitions in greater detail.

Light is emitted and absorbed by atoms during their transition from one energy state to another. This is characteristic not only of atoms, but of molecules and ions as well. The process of the particle transfer from its normal (stationary) state corresponding to the minimum energy of the system to a higher energy state is termed *excitation* and the particle itself is said to be *excited* (Fig. 5a). This process is accompanied by the absorption of the energy of the external field.

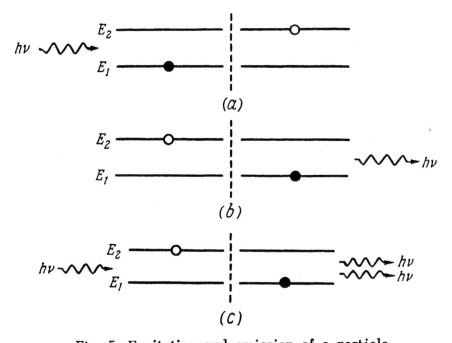

Fig. 5. Excitation and emission of a particle

(a) excitation of a particle accompanied by absorption of a photon. The energy of the particle as a result of excitation corresponds to a higher energy level; (b) spontaneous radiation accompanied by the origination of a photon. The energy and frequency of the photon are determined by the energy difference in the levels between which the transition took place; (c) induced radiation. As a result, two photons originate, whose energy, frequency and phase are the same

The particle (an atom, molecule or ion) can pass to an excited state not only when it absorbs a quantum of the electromagnetic field but also when colliding with other particles of the same kind (atoms, molecules or ions) that have a certain store of energy. The system can be excited, e.g. if a flux of electrons or electric current is passed through it.

Usually the number of excited atoms in a system, i.e. of the atoms whose energy corresponds to a higher energy level, is smaller than the number of non-excited ones. The time during which an atom can exist in its ground state is unlimited. On the contrary, in an excited state an atom can remain only for a limited period of time which is termed lifetime and is denoted by τ. For example, the lifetime τ of excited hydrogen atoms is of the order of 10^{-8} s. However, there exist such excited states which are characterized by a relatively long lifetime ($\tau \gg 10^{-8}$ s). These states are called *metastable*.

The transition of an atom from one energy level to another can also be non-radiative. In such a case energy is transmitted to some other atom and converted into heat.

Only certain transitions are possible in the atom, which are determined by the probability value and allowed by *selection rules*. The set of allowed transitions between the energy levels makes up the *energy spectrum of an atom*. This spectrum consists of series of lines separated by forbidden intervals.

For an atom which is in an excited (higher energy) state there exists a probability that after some time it will return to its ground (lower energy) state. For the quantitative estimation of the number of transitions possible in one atom of an excited system per second, the concept of *transition probability A* is resorted to. The average transition probability for a large number of similar atoms has a strictly definite value. The transition probability A and the average lifetime τ of a given energy level are reciprocal quantities

$$A = 1/\tau$$

When passing from one energy state to another, an atom radiates energy. The process of radiation when the transition of an atom to a lower energy level is spontaneous (not caused by any extraneous effects) is called *spontaneous radiation* (Fig. 5b).

Spontaneous radiation is specific in being of a random character, since such radiation is a random mixture of quanta having various wavelengths. The waves coincide neither in their length nor in phase. This kind of radiation is therefore incoherent and has a broad spectrum.

What the components of radiation of a conventional incoherent light source are like can be seen in Fig. 6

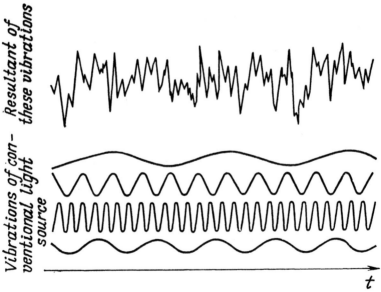

Fig. 6. Incoherent radiation

(bottom part of it). The length and amplitude of the waves radiated by such a source can be quite diverse. The difference in the phases of oscillations is also random in character. In the top part of the same figure the resultant curve equal to the sum of individual oscillations is shown. The character of this curve is complicated, and it would be very difficult to find any periodicity in the amplitude variation.

In 1917, Einstein predicted that besides spontaneous emission there must exist still another, *induced radiation*. Sometimes such radiation is called stimulated. This implies that a particle (an atom, molecule or ion) can pass from an excited state to its normal state emitting a light quantum (photon) not only spontaneously, but also when forced to it, under the effect of another external quantum. Induced radiation is a process opposite to absorption. Contrary to absorption during which a light quantum disappears, induced radiation is associated with the appearance of a new quantum (see Fig. 5c).

Quanta of electromagnetic radiation, which owe their origin to the effect of an external magnetic field are absolutely indistinguishable from those light quanta (photons) which have caused the atom to pass to a lower energy level. The quantum of electromagnetic energy, which induces stimulated emission undergoes no alterations either. A photon, having encountered an excited atom in its path, "knocks out" a similar photon from the latter. The wavelength, direction of propagation and phase of the both photons are in a strict coincidence. The resulting radiation is coherent and the spectral line is narrow.

The character of coherent radiation can be better understood when considering Fig. 7. In the bottom part of this figure you can see the diagrammatic presentation of oscillations that are equal in frequency and

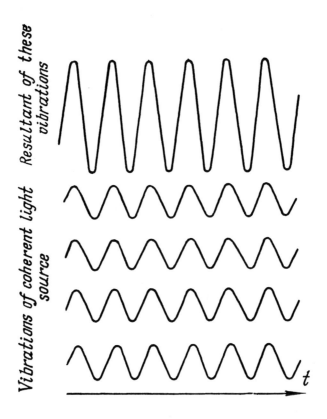

Fig. 7. Coherent radiation

have the same initial phase. The phase difference remains constant for a very long period of time. Since the phases of all these oscillations are of the same sign, they coincide and their amplitudes are added together. No mutual damping of the oscillations takes place. The intensity of the resultant oscillation (shown in the top part of the figure) equal to the sum of all the components is in this case very high.

Induced transition leads to the appearance of two photons instead of one; in other words, their number is doubled. Now, if a system is in an excited state, these two photons may cause further induced transitions, and so on. The result will be an avalanche process. Ultimately, the light which has passed through such an excited medium will be amplified. For this reason an excited medium is called an *active medium*. The above situation, however, is possible only when the majority (more than half) of all the particles in a system are in an excited state. Otherwise the process will be attenuating and no amplification of the light passing through the given medium will take place. Evidently, amplification of light (generation) can be attained by specially creating such active media. As has been pointed out above, under normal conditions the number of excited particles even in an excited system is always less than the number of non-excited ones. For the number of excited particles in a system to exceed the number of particles in the stationary state, special conditions are required.

DISTRIBUTION OF PARTICLES AMONG ENERGY LEVELS

In any substance the atoms can perform not only spontaneous transitions from a higher energy level to a lower one, but also direct and inverse transitions under the effect of a thermal radiation field. As a whole, the substance is in the state of thermodynamic equilibrium, since the probability of these direct and inverse transitions is the same. But the number of particles which are found in an excited state depends on the temperature. The equilibrium distribution of particles among energy levels is described by the well-known Boltzmann formula

$$N_i = N e^{-\frac{E_i}{kT}}$$

where N_i is the number of particles in the state of energy E_i

N is the total number of particles:

k is Boltzmann's constant equal to $1.38 \cdot 10^{-16}$ erg/deg; and

T is the absolute temperature

It will be recalled here that the absolute scale of temperature (Kelvin's) is related to the Celsius scale by the equation

$$T = t° + 273.16°C$$

The zero point of Kelvin's absolute scale of temperature corresponds to the temperature $t° = -273.16°C$.

As can be seen from the Boltzmann formula, the number of excited particles depends, first of all, on temperature. The ratio between the number of atoms in two definite energy levels, or the population of these levels, is given by the equation which can be deduced from the Boltzmann formula

$$\frac{N_2}{N_1} = e^{-\frac{E_2 - E_1}{kT}}$$

From this formula it follows that with any positive T the number of particles in a higher energy level decreases with a higher serial number of this level; there holds the following inequality:

$$\text{for } E_1 < E_2, \quad N_1 > N_2$$

In other words, the population of higher energy levels is less than the population of lower energy levels. A system with such distribution of particles absorbs quanta of electromagnetic energy.

The creation of systems in which the number of particles in a higher energy level is greater than in a lower energy level turned out to be practically possible. These systems are termed *inverted population systems.* If we refer once again to the Boltzmann formula, we shall see that this situation corresponds to a state when the absolute temperature is negative (i.e. when the exponent in the formula is positive). The states of a system, in which the population of a higher energy level exceeds that of lower energy level are called *negative temperature states.* It should be borne in mind that here negative temperature is to be understood not as a physical quantity, but only as a convenient mathematical expression signifying the non-equilibrium state of the system.

If a system has several energy levels, one of them may have a negative temperature with respect to some other level (or levels), whereas with respect to the remaining levels its temperature will be positive.

The state of a system having a negative temperature is unstable. Spontaneous radiation, an external electromagnetic field return the system to its steady state. Hence, there appears a possibility of mass induced transitions in the system and, consequently, of attaining an amplification of the electromagnetic radiation.

But in what way can a negative temperature state be obtained? What is to be done in order that the population in the higher energy levels of an active medium should exceed the population in its lower energy levels?

ACTIVE SYSTEMS

Suppose there is a system of atoms (molecules or ions) capable of having only two energy levels (states), namely, the lower level E_1 which corresponds to the stationary, non-excited state and the higher level E_2 which corresponds to the excited state.

If we excite this system by using an electromagnetic field that has a frequency corresponding to the difference of the transitions

$$\nu = \frac{E_2 - E_1}{h}$$

some particles of the system will pass from the lower to the higher level. At the same time induced transitions, i.e. those from the higher to the lower level, will also take place. These transitions will be accompanied by the emission of quanta of electromagnetic energy. If the exciting field intensity is sufficient, after a certain period of irradiation one half of the atoms in the irradiated system will be in the excited state and the other half, in the stationary state. Since the probability of induced radiation is the same as that of excitation, from a certain moment of time the number of the atoms passing from the higher to the lower level will be equal to the number of the atoms rising to the higher level. A dynamic equilibrium will thus be established in the system.

If now a procedure could be found for separating the excited atoms, in other words, for creating an active medium in which the excited atoms would prevail, then the radiation passed through such medium with the frequency corresponding to the transition will be amplified by it.

Such a method of radiation amplification has found practical application in molecular amplifiers (oscillators) where molecules of ammonia preheated in a special furnace are separated. Molecules of ammonia, which after such heat treatment acquire different energy states are then passed through a non-uniform electric field created by specially designed (quadrupole) capacitors. When a bunch of ammonia molecules is passed through such a capacitor, excited molecules will be concentrated mainly along the capacitor axis, while non-excited molecules will be deflected towards its periphery.

Such a bunch of excited molecules that possess excess energy is inversely populated. If a radiation with an appropriate wavelength ($\lambda=1.27$ cm) is passed along such bunch, this radiation will be amplified. In practice this effect is attained in special cavity resonators. These are so-called two-level systems.

However, the method of obtaining a negative temperature medium by using two-level systems is not free from serious disadvantages. Systems with two energy levels, for instance, amplifiers based on this method are capable of operating only at one frequency.

A more successful method for the excitation of quantum systems and producing a negative temperature medium, based on the use of three-level systems, was suggested by N. Basov and A. Prokhorov.

Suppose we have a system possessing three energy levels, and all transitions between these levels are allowed. If we expose this system to the effect of a field from an external source (oscillator) with a frequency

$$\nu = \frac{E_3 - E_1}{h}$$

corresponding to the transition from level *1* to level *3*, some of the particles will pass to level *3*.

In case the irradiation intensity is higher, the number of particles in the higher level *3* and in the lower level *1* will be the same ($N_3 \approx N_1$), that is, the quantum transition from E_3 to E_1 will be saturated (Fig. 8a).

Provided the lifetime of the particles in level *3* is sufficiently long, their number in this level will exceed that in level *2*; in other words, level *3* will be inversely populated with respect to level *2*. This will be an active system.

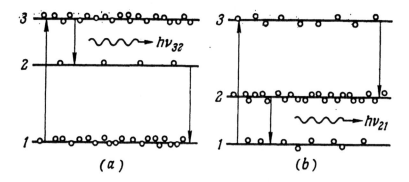

Fig. 8. Diagram of producing inverted population in three-level systems

(a) inverted population between energy levels *3* and *2*; (b) inverted population between levels *2* and *1*

Now, if through this system we pass an electromagnetic radiation with a frequency

$$\nu = \frac{E_3 - E_2}{h}$$

corresponding to the transition from level *3* to level *2*, this radiation will be amplified due to mass induced transitions from level *3* to level *2*. The amplification will take place as a result of liberation of the excess energy the system acquired during its initial excitation with the electromagnetic field from the external source. Mass induced transitions in an excited system of such kind may also be caused by random spontaneous transitions. In this case the system will behave as a generator.

From level *2* the particles then return to level *1*.

It is also possible to create systems in which excited particles will accumulate in level *2* and mass induced transitions will occur between levels *2* and *1*; the wave that has initiated the induced transition and that corresponds to the transition between these two levels will be amplified (Fig. 8*b*).

Besides three-level systems there also exist and

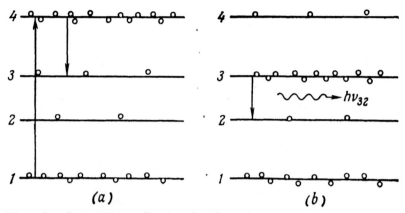

Fig. 9. Operation diagram of a four-level system
(a) population of levels at the first moment after excitation; (b) population of levels by the moment of induced radiation transitions

find practical application four-level systems. Most of these systems operate on the following principle. Under the effect of radiation emitted by an excitation source, particles of an active substance pass from energy level *1* to energy level *4* (Fig. 9a). Energy level *3* is metastable. The transition of electrons from level *4* to level *3* is non-radiative. The inversion of populations is created between levels *3* and *2* (Fig. 9b). Level *2* is sufficiently remote from level *1* and, therefore, under thermodynamic equilibrium conditions, its population is close to zero. Consequently, for obtaining a negative temperature state in a four-level system a considerably less powerful excitation is required than in a three-level system where the inverted population is obtained with respect to level *1* which, being the ground level, usually remains considerably populated.

These methods of creating negative temperatures and inverted population levels by using three- and four-level systems have found wide application in optical quantum generators (lasers) and are fundamental for their operation.

CHAPTER 3

Generators of Light

RUBY LASER

Let us consider the operation of an optical quantum generator employing a ruby crystal as the working material. Such a generator of light is called a *ruby laser*. Ruby lasers are most widespread type of generators using solid crystalline substances as their active material.

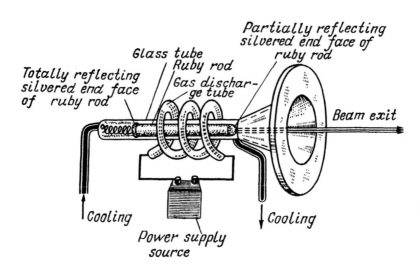

Fig. 10. Design of a ruby laser

This laser consists of three main parts: an active (working) material, a resonant system made as two parallel plates with reflecting coatings applied on them, and an exciting system usually made up of a helical xenon flash tube and a power supply source (Fig. 10).

Ruby is a crystal of aluminium oxide where part of aluminium atoms are substituted by chromium atoms ($Al_2O_3 : Cr_2O_3$). The active material in the ruby are chromium ions Cr^{3+}. The colour of a ruby crystal depends on the content of chromium in it. For lasers use is usually made of pale pink ruby crystals containing about 0.05 per cent of chromium. Ruby crystals are grown in special furnaces, then annealed and shaped into rods. Such rods are 2 to 30 cm in length and from 0.5 to 2 cm in diameter. Flat end faces of the rod are made strictly parallel, ground and polished to a high degree of precision. Sometimes reflecting coatings are applied not on special plates but directly on the end faces of the ruby rod. The end faces of the rod are silvered so that the surface of the one end face becomes fully reflecting and that of the other end face, partially reflecting. Usually the light-transmission coefficient of the partially reflecting end face of the ruby rod is about 10 to 25 per cent, but other values are also possible.

The ruby rod is arranged along the axis of a helical xenon flash tube in such a manner that the coils of the helix encompass the rod. The flash of the tube lasts several milliseconds. During this period of time the tube consumes energy amounting to several thousand joules and most of it is spent for heating the apparatus. The other smaller part of the energy in the form of blue and green radiation is absorbed by the ruby. This energy ensures the excitation of chrom˙ ions.

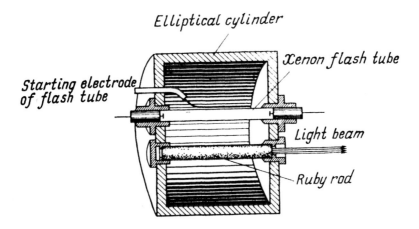

Fig. 11. Design of a laser with a ruby rod and a flash tube housed in an elliptical reflector

Fig. 11 shows a laser of a different design. In this laser a flash tube is straight. It is placed in one of the focal axes of an elliptical reflecting cylinder, while a ruby rod is in the other focal axis of this cylinder. A pulse power supply for the xenon flash tube is obtained from a high-capacity capacitor charged to a required voltage from a rectifier.

An energy diagram illustrating the operation principle of a ruby laser is shown in Fig. 12. In this diagram lines *1*, *2* and *3* correspond to energy levels of

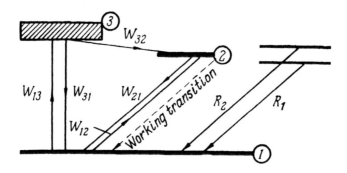

Fig. 12. Schematic diagram of energy levels in ruby

chromium ions. Negative temperatures (inverted population) in a ruby laser are obtained by using a three-level system.

In the normal, non-excited state chromium ions are in the lower level *1*. When the ruby crystal is irradiated with the light of a xenon flash tube, containing the green component of the spectrum, the chromium atoms are excited and pass to the upper level *3* where the light absorption band is 5600 Å. The absorption band width of this level is about 800 Å.

From level *3* part of the excited chromium atoms return to the ground level *1* and the other part, to level *2*. The so-called non-radiative transition takes place, during which chromium ions give off part of their energy to the crystal lattice in the form of heat. The probability of transition from level *3* to level *2* is 200 times greater and from level *2* to level *1* 300 times smaller than the probability of transition from level *3* to level *1*. As a result, level *2* turns out to be more populated than level *1*. In other words, the population is inverted and thus conditions required for intensive induced transitions are created.

As we know, such a system is extremely unstable. The probability of spontaneous transitions at any moment of time is very high. The very first photon appearing during such spontaneous transition, in accordance with the law of induced emission, will knock out a second photon from a neighbouring atom and the atom from which the first photon was emitted will be brought to its ground state. Now these two photons will knock out two more photons and their total number will be four, and so on. The process grows practically instantaneously. The first wave of radiation, on reaching the reflecting surface, will return and cause further increase in the number of induced transitions and in the radiation intensity.

Such a process will repeat many times. The generation will rise and the power will increase till the majority of the excited particles of the active material (chromium ions) give off the energy acquired at the moment of excitation. All this will take place on condition that power losses at reflection (due to imperfections of the reflecting coatings and to that one of the end faces of the rod is made semi-transparent so that a radiation flux will start emerging from it at the very beginning of the generation) do not exceed the power acquired by the beam forming in the laser rod as a result of the commenced generation. A very high-intensity beam will emerge through the partially silvered end face of the ruby rod. The direction of this beam will be strictly parallel to the ruby axis (Fig. 13).

Those photons the direction of propagation of which at the moment of origination failed to coincide with the axis of the ruby rod will leave the rod through its side walls without having caused any noticeable generation.

It is just the multiple pass of the resulting light wave between the end walls of the resonator without any substantial deviation from the ruby rod axis that ensures strict directivity and tremendous output power of the laser beam.

Since level *2* (Fig. 12) is in fact constituted by two close sub-levels, in case the exciting power is insufficient, two weak lines R_1 and R_2 are emitted, their respective wavelengths being 6943 and 6929 Å (Fig. 14a). The width of these lines is about 6 Å. This radiation is mainly due to spontaneous transitions.

With an increase in the exciting power, the intensity of radiation at the wavelength of 6929 Å is practically no more increasing, and at the wavelength of 6943 Å

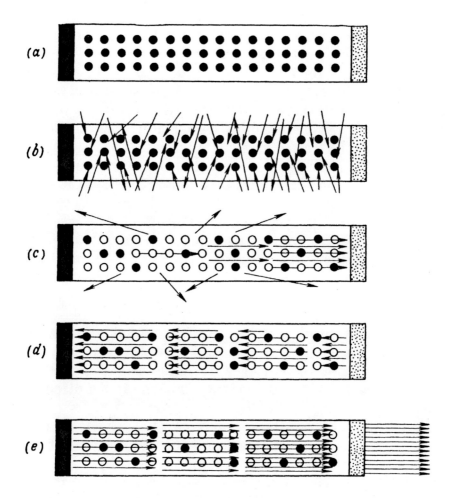

Fig. 13. Process of shaping a beam in the active medium of a laser

(*a*) atoms of the active medium in non-excited state; (*b*) pumping light transfers most of the atoms to the excited state; (*c*) some of the atoms radiate spontaneously: part of the photons rush outside, some photons, moving parallel to the rod axis, cause induced radiation; (*d*) having reflected from the mirror surface of the rod end face, the flux of photons is amplified while passing through the excited medium; (*e*) light beam emerges through the partially silvered surface of the ruby rod end face

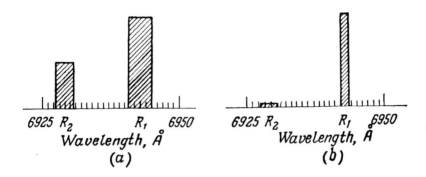

Fig. 14. Emission spectra of a ruby laser
(a) emission spectrum in case of insufficient pumping power; (b) emission spectrum at high pumping power

generation is established, the radiation drastically increases and the spectral line become narrower (Fig. 14b). Most ruby lasers operate in a pulse mode. The emitted pulse usually is produced some 300 μs after the excitation pulse, this time lag being necessary for the creation of a difference in the population of the levels. The duration of a laser pulse is several milliseconds and the energy of a single pulse reaches dozens and hundreds of joules.

The emissive power of a laser grows with an increase in the concentration of active particles and the dimensions of its active element. But the laser power cannot be raised indefinitely, limitations being placed by internal losses which are the higher the greater the dimensions of the active element.

Studies of pulses emitted by the laser have shown that each such pulse consists of several hundreds of short pulses called "spikes" with a duration of about 1 μs and a rise time of 0.1 μs. These pulses are superposed on a certain average level of radiated power. Intervals between these "spikes" are of the order of 5 to 10 μs (Fig. 15).

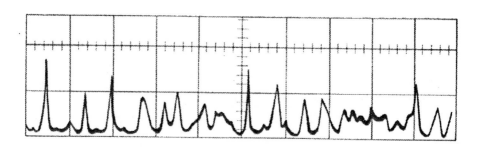

Fig. 15. Radiation of a ruby laser on an expanded time scale (1 cm corresponds to 5 μs)

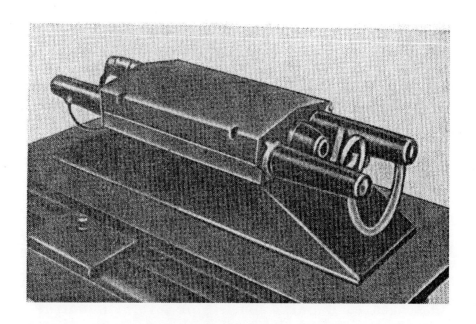

Fig. 16. Soviet ruby laser

In some cases it is desirable that pulse emission should be stable. To this end attempts have been made at creating stabilising systems which could ensure sufficiently stable emission instead of random pulses, smaller width of the spectral line, smaller divergence angle and greater coherence of radiation.

Figure 16 is a photograph of one of the ruby lasers made in the Soviet Union. The ruby rod of this laser is 240 mm long and has a diameter of 12 mm. The period of one laser pulse is several minutes and its energy is 40 to 60 J. The laser is intended for studying the processes of interaction between high-intensity light and various substances.

PROPERTIES OF LASER BEAM

An extremely important feature of laser radiation is its coherence. A high coherence of the laser beam can be verified by a procedure known as Young's experiment on interference.

This experiment requires the use of a remote source of light having a small size. Such a source can be replaced by a slit illuminated by a remote light source (this experiment will be difficult to carry out if the light source is either too large or located near the screen). If we pass a light pencil from a remote source through two narrow parallel slits, then an interference pattern of alternating bright and dark bands will be observed on a screen located behind the slits. Now, if instead of a remote source of light we take a laser and make two slits directly on the semi-transparent silvered surface of one of its end faces through which radiation will emerge, a very distinct interference picture will be easily obtained on the screen (Fig. 17). This experiment confirms that different

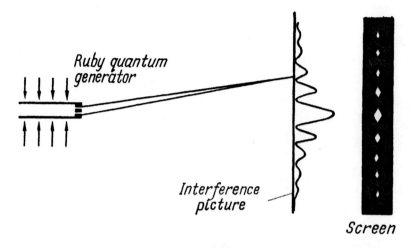

Fig. 17. Experiment on interference, confirming coherence of laser radiation

points of radiation emerging from the laser are coherent.

Another important property of laser radiation is its emission in the form of a narrow directed beam.

In the general case the width of the beam of an optical quantum generator is determined by diffraction and the diameter of the rod of active material

$$\theta = 1.22 \frac{\lambda}{D} \text{ (radian)}$$

where θ is the beam angle, radians; λ is the radiation wavelength, cm; D is the diameter of the ruby rod, cm.

Hence, the shorter the wavelength and the larger the rod diameter, the smaller the angular divergence of the laser beam.

However, an increase in the rod diameter brings about a sharp increase in the number of vibrations in the resonant cavity, with a frequency differing from the working frequency of the laser. These vibrations propagate within the resonant cavity at larger angles to the rod axis than the vibrations at the working frequency. As a result, when the diameter of the active material is increased, the monochromaticity and coherence of laser emission become deteriorated and therefore the possibility of narrowing the beam to the theoretical limit is rendered difficult. According to calculations, the width of a laser beam in an arrangement employing a ruby rod of 1 cm in diameter should not exceed 10^{-4} radian. Practically, however, the beam divergence exceeds this theoretical value. For example, for a ruby crystal of 1 cm in diameter the beam divergence is $6 \cdot 10^{-3}$ radian. Under thoroughly controlled conditions it can be brought down to $1 \cdot 10^{-3}$ radian.

Nevertheless, high coherence and monochromaticity of laser emission allow the light beam emerging from the laser to be focused with the help of a system of lenses onto an extremely small surface area. In the limit, the diameter of the spot of the focused laser beam approaches the length of the wave radiated by the laser. If we take a conventional light source (such as a common electric lamp), our attempt at focusing its light rays into one point will prove unsuccessful, because these rays are not parallel and propagate in all directions. In the focal plane we shall obtain an image of the object (light source) and, however small the dimensions of this image may be, they will be far greater than a geometrical point (Fig. 18a). Another reason why our attempt will fail is that radiation from a conventional light source contains a great number of components with different wavelengths and these components undergo different refractions in the focusing lens. Consequently, only radiation of one particular wavelength can be focused in the focal plane (Fig. 18b). Radiation emitted by the laser is, on the contrary, strictly parallel, has only one definite and

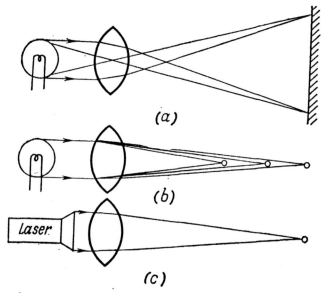

Fig. 18. Conventional light source and laser beam characteristics

(a) light beam cannot be made to converge at one point on account of its separate components being non-parallel; the light source has finite dimensions; (b) light beam cannot be made to converge at one point on account of different wavelength of its separate components; (c) light beam of a laser can be made to converge at one point commensurate with the radiation wavelength because laser beams are parallel and monochromatic

constant wavelength, and therefore, with the help of optical systems, it can be focused into one point, the result being a spot having a diameter close to the radiation wavelength (Fig. 18c). This circumstance makes possible the obtaining of an extremely high power density of laser radiation.

The minimum diameter of the light spot for a ruby laser is 0.7 μ. In the general case the spot diameter is determined by the focal length of the lens and angular divergence of the beam. If these values are known, the possible minimum diameter of the spot can easily be determined from the following simple formula:

$$d = f\theta$$

where f is the focal length of the lens, m; d is the spot diameter, m; and θ is the angular divergence of the beam, radian.

For example, in case the focal length of the lens is 5 cm and the angular divergence of the beam is chosen to be 10^{-4} radian, then

$$d = 0.05 \cdot 10^{-4} = 5\,\mu$$

The shorter the focal length, the smaller the dimensions of the light spot will be.

If we assume that the power of radiation generated by a laser is one million watts and the beam diameter at the laser outlet is 1 cm, then, if this power is concentrated on a surface area of 5 μ in diameter, we shall obtain the luminous flux density of $4 \cdot 10^{12}$ W/cm²! So fantastic a density of the luminous flux, however, will be obtained only in the immediate vicinity of the focal point. Any extended thin pencil of light cannot be obtained on account of diffraction.

Since the laser beam exhibits neither ideal coherence nor ideal monochromaticity, the spot diameter actually proves to be 100 to 150 times larger, and the power density of the luminous flux diminishes accordingly. But there are sound grounds to expect that with advances in crystal growing and crystal finishing techniques and better qualities of mirrors and optical systems the power density of the luminous flux will be considerably increased.

The pulses of the so far created lasers have a duration of 10^{-8} s and energy of several dozens of joules. This means that the instantaneous power in a pulse reaches one thousand million watts. When focused into a small spot, such a beam creates the flux density lying within the above-considered range: 10^{12} W/cm² and even higher. A few years ago scientists could only dream of such a colossal concentration of power.

It will be of interest to consider the character of the processes which take place at the point of the laser beam focusing. The character of these processes is obvious to depend on the beam power. With the flux density of the order of 10^6 to 10^8 W/cm^2, there will take place an intensive evaporation from the surface of a sample exposed to the effect of the beam. The temperature on the surface of the sample reaches several thousand degrees. For instance, one pulse with an energy of about 10 J and a duration of one millisecond can shoot through a 3 mm thick plate. The hole made in the plate has a small convergence angle, of the order of 3 to 4°, and the length of the hole is 5 to 15 times its diameter. Vapours of the substance have time to scatter without screening the crater.

If the density of the luminous flux is increased to 10^8 W/cm^2 and the duration of the pulse is shortened to 0.01 µs, no hole will be made in the plate, but a fused spot will appear at the point where the beam was focused. Plasma that has formed in this place, probably has no time to dissipate, absorbs all the energy of the light beam and then, when dissipating, transmits a mechanical impulse to the surface of the substance. Experiments with thin foils confirm such an assumption: a hole is made in the foil, with its edges pressed out as if by a shock wave. At higher powers a plasma breakdown occurs, accompanied by a cracking sound. In case the power is increased still further, a plasma breakdown will take place already in air.

Such is, in general terms, the picture of the phenomena taking place in a focused laser beam.

ACTIVE MATERIALS

Active materials most often used in lasers are substances having a crystalline structure. At the time when first lasers were created such substances were best studied. It is not by chance either that the first active material employed in a laser was artificial (synthesized) ruby. Corundum (and ruby is a variety of corundum coloured due to the presence of chromium atoms in it) has found wide application in most diverse branches of engineering, which fact, no doubt, contributed to advances in the technology of growing corundum crystals. The colour of ruby depends on the percentage of chromium in it: the higher this percentage, the more intensive the colour of ruby is. Optimal percentage of chromium in ruby is considered to be 0.05. But rubies with other concentrations of chromium ranging from 0.005 to 0.5 per cent are used as well.

The red colour of ruby is accounted for by that chromium atoms in the crystal absorb light in the broad green, blue and indigo bands of the visible spectrum,

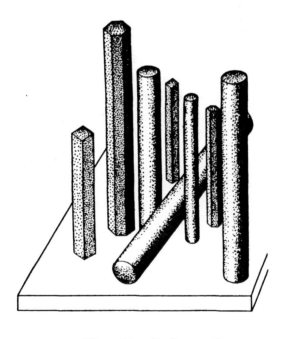

Fig. 19. Ruby rods

as well as in the ultra-violet region, passing the light in the red region only. Ruby crystals are manufactured into rods that are from 2 to 30 cm long, have a diameter from several millimetres to 1-2 cm and are round, hexagonal or square in cross-section (Fig. 19). Rods with round cross-sections are preferred since they are less difficult to manufacture.

Experimental data which have been accumulated allow formulation of definite requirements on ruby rods. Thus, the deviation of the optical axis of the crystal from the geometrical axis of the rod should not exceed 5'. The end faces of the rod should be worked with a particular precision. Permissible deviation of the shape of the surface of the rod end faces from the plane is 0.1 of the wavelength at which the laser is to operate. Particular care is taken that the end faces of the rod should be parallel to each other within 2", and that the angle between the plane of the end face and the cylinder generatrix should not deviate from 90° by more than 1'.

All the above-listed requirements pursue the objective of improving the beam characteristics and obtaining the generation with a minimum possible power of the excitation source.

The shape and dimensions of crystal rods are selected depending on the required power of radiation and are also determined by the characteristics of the source and optics of the excitation system, the design of the rod holder and the method of cooling adopted.

The quest for a higher laser-conversion efficiency and for obtaining stimulated emission at new frequencies of the optical region necessitates the study of other crystalline materials. For this purpose crystals of various halides, tungstates, titanates, molybdates and other materials with addition of rare-earth and other elements as activators are grown.

Besides the wide-spread ruby laser whose design and operation principle have been discussed above, lasers employing other crystalline substances are also used. Most of them operate on the principle of a four-level system (Fig. 20). Characteristic of a four-level system is that induced radiation originates when the atoms of an active material pass from their metastable state not to the ground level as is the case with a three-level system but to a certain intermediate level. The population of this intermediate level at a low temperature proves to be not high, and the power required for exciting the system is less than that needed for three-level laser systems. An increase in the working temperature drastically reduces the laser effectiveness; such lasers yield adequate results only at low temperatures and most of them cannot operate at room temperature.

Lasers on uranium-doped crystals. Calcium fluoride doped with trivalent ions of uranium ($CaF_2:U^{3+}$) was one of the first crystals employed in lasers.

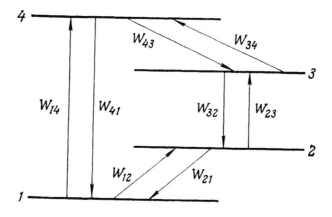

Fig. 20. Scheme of transitions in a four-level system

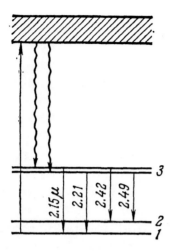

Fig. 21. The energy level scheme of uranium

A simplified diagram of energy levels in such a crystal is shown in Fig. 21. Ions of trivalent uranium display several absorption bands in the visible and near infra-red regions of the spectrum. The strongest absorption bands lie in the green and indigo parts of the spectrum. Intensive emission lines lie in the infra-red region.

At room temperature the emission spectrum of calcium fluoride doped with uranium consists of four lines, two of which coincide with the absorption lines and correspond to the transition from metastable state *3* to ground state *1*. The two other emission lines are associated with the transition to intermediate level *2*. Induced radiation takes place as a result of transition from the metastable state *3* to state *2* at a wavelength of 2.49 μ.

The arrangement of such a laser is the same as of a ruby laser. This laser operates at a temperature of liquid helium. A xenon flash tube can be used in it as an excitation source.

Lasers on neodymium-doped crystals. A crystal of calcium tungstate doped with trivalent ions of neodymium can be used as the active material in such lasers.

The energy diagram in a crystal of calcium tungstate is rather complicated and therefore we shall not give it here. The absorption spectrum of neodymium ions lies mainly in the visible and near infra-red regions. Neodymium ions have several emission lines. One of them, which is the strongest at the temperature of liquid nitrogen, has a wavelength of 1.063 μ. In a pulse mode, a laser employing such a crystal can operate both at room temperature and at the temperature of liquid nitrogen. At room temperature the emission wavelength is 1.0646 μ. A xenon flash tube is also used for excitation.

Laser crystals can be doped with other rare-earth elements as well, for instance, with bivalent samarium included into crystals of calcium fluoride ($CaF_2:Sm^{2+}$). Induced emission of samarium ions takes place at a wavelength of 7080 Å at temperatures of liquid hydrogen and liquid helium. Besides crystals of calcium fluoride doped with bivalent samarium ions, crystals of strontium fluoride with the same dopant are employed: $SrF_2:Sm^{2+}$. The wavelength in this case is 6969 Å. For lasers operating in the infra-red region use can be made of trivalent ions of thulium and holmium included into crystals of calcium tungstate.

For CW lasers the use of calcium tungstate crystals doped with trivalent praseodymium ions ($CaWO_4:Pr^{3+}$) is rather promising. The induced radiation wavelength of this material is 10 470 Å and the excitation power is 30 J.

For ensuring CW operation of a laser, crystals of calcium fluoride doped with bivalent dysprosium ions ($CaF_2:Dy^{2+}$) are employed. This material features a very low excitation level (of the order of 0.1 to 1.0 J) since it absorbs wide frequency bands. The induced radiation wavelength is 23 600 Å.

The number of rare-earth elements which can find application in lasers is rapidly increasing. The range of wavelengths at which stimulated emission can be obtained is also extending. Crystals doped with these elements are also produced in the shape of small rods having a length of several centimetres.

Table 2

Main Characteristics of Solid Active Media Employed in Lasers

Active medium		Emission wavelength λ, Å	Working temperature, °K	Laser operation
activator	matrix			
Chromium, Cr^{3+}	Al_2O_3	6 929	300	Pulsed
		6 934	77	CW
		6 943	293	Pulsed, CW
Uranium, U^{3+}	CaF_2	25 100	300/77	Pulsed, CW
		26 130	300/77	Pulsed, CW
Ditto	SrF_2	24 070	90	Pulsed
Ditto	BaF_2	25 560	20	Pulsed
Samarium, Sm^{2+}	CaF_2	7 082	20	Pulsed
	SrF_2	6 969	4	Pulsed
Dysprosium, Dy^{2+}	CaF_2	23 600	293/27	Pulsed, CW
Neodymium, Nd^{3+}	CaF_2	10 461	300	Pulsed
	SrF_2	10 370	293	Pulsed
	BaF_2	10 600	77	Pulsed
	$CaWO_4$	10 646	293	Pulsed
		10 650	77	CW
	$CrMoO_4$	10 643	293	Pulsed
Thulium, Tm^{3+}	SrF_2	19 100	77	Pulsed
	$CaWO_4$	19 110	77	Pulsed
Praseodymium, Pr^{3+}	$CaWO_4$	10 468	90	Pulsed
	$SrMoO_4$	10 470	77	Pulsed

For attaining a higher effectiveness of stimulated emission, in some cases coatings are applied on the side surface of crystals. For example, ruby crystals with a sapphire coating are grown. These crystals are single rod-shaped structures. A ruby rod and a sapphire tube for such structures can be manufactured separately with a high degree of precision. Then the rod is inserted into the tube so that a close contact between them is ensured. The sapphire tube functions as a refracting envelope enhancing the concentration of light incident on the ruby core and, at the same time, contributing to a more rapid cooling of this core.

A very important factor which determines the character of the emerging radiation is the orientation of the crystal, i.e. the relative position of the optic axis of the crystal and of the rod axis. If the optic axis of the crystal and the rod axis are parallel, the orientation is said to be zero. Rods with the zero orientation give a circular or an elliptical polarisation of the beam. Rods with a 90° orientation (when the optic axis is perpendicular to the rod axis) give beams polarised in one direction. Rods with other orientations of the optic axis, e.g. with that of 60°, are also used.

The main characteristics of crystal lasers are given in Table 2.

METHODS AND SOURCES OF EXCITATION

Excitation of an active substance in lasers is often termed pumping. There are several methods of pumping. Most widespread and simplest of them is optical pumping.

Most conventional sources of optical pumping are flash-discharge tubes, mainly xenon ones. Such tubes may vary in design. The pumping system usually

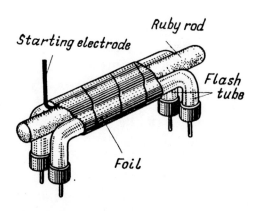

Fig. 22. One of possible laser excitation systems. The ruby rod together with two U-shaped flash tubes is wound with a layer of foil functioning as a reflector

consists of two main elements: a light source and a reflector which concentrates the light from the source on the active material.

The character of laser emission is closely related to the pumping power. When this power is small, the emitted pulses contain a great number of short spikes which last for 1 µs and are separated by 5 to 10 µs intervals. With the pumping power increasing, the intervals between the spikes become shorter and the amplitude of the spikes rises.

The first models of ruby lasers employed helical flash-discharge tubes encompassing the ruby rod (Fig. 10). The energy provided by such a pumping system was approximately 2000 J. Later, other more effective pumping systems were developed.

In one of such systems a ruby rod is positioned between two flash tubes which almost touch the rod. A layer of foil which functions as a reflecting cover is wound around the flash tubes and the crystal rod (Fig. 22). The pumping threshold for this arrangement is 120 to 200 J. Sometimes for increasing the pum-

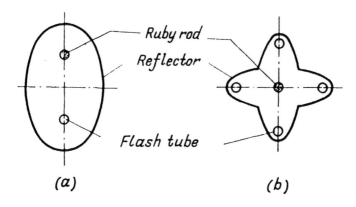

Fig. 23. Types of reflectors
(a) elliptical; (b) with four elliptical reflectors

ping power four or six flash tubes are arranged in a similar manner in the pumping system.

Another design uses an elliptical cylindrical reflector housing a ruby rod disposed along one focus line of the reflector and a pumping flash tube disposed along the other focus line of the reflector (Fig. 23a). Such a design ensures an effective utilisation of the flash tube power, since its entire radiation is focused on the ruby rod. The threshold energy in this case comes to several hundreds of joules.

In those cases when a high output power is to be obtained, a pumping system with several, e.g. with four reflectors can be used. Such a system is shown schematically in Fig. 23b. In this arrangement four elliptical reflectors have a common focus line along which a ruby rod is placed. Four flash tubes are placed at the other respective focus lines of the elliptical cylinders. This provides a possibility of focusing the light from the four flash tubes on the ruby rod. With the dimensions of the reflector system appropriately selected, the efficiency of each of the four reflectors will be about 75 per cent that of a system using one elliptical cylinder. Lowering of the efficiency is associated with a reduction of the focusing surface area. The structure consisting of four elliptical cylinders appears to be optimal. Though a greater number of these cylinders will lead to a higher pumping power due to the addition of new flash tubes, yet there will be a reduction in the degree of energy concentration by individual reflectors since their reflecting surface areas diminish. The best conditions for a complete utilisation of the energy radiated by the flash tube in such a reflector are the largest possible cross section of the elliptical reflector, minimum diameter of the flash tube and minimum diameter of the rod of active material.

With this kind of a system, the total pumping energy is 8000 J, the duration of the flash tube pulse is 1 µs, and that of the laser pulse is 0.5 µs.

The threshold pumping power can be reduced by using a ruby rod with a sapphire envelope (Fig. 24) which focuses the pumping light in such a manner that the radiation intensity is concentrated in the middle portion of the ruby rod (Fig. 25) where optical excitation is most easily set up. The light pumping energy which passes along the sapphire layer towards the centre of the ruby rod is absorbed to a considerably smaller extent than would be the case with the same layer of ruby instead of sapphire. Besides, sapphire is a good conductor of heat, and this contributes to rapid cooling of the ruby. Ruby rods with a sapphire envelope are excited with a pumping energy of 50 to 80 J.

The repetition frequency of excitation pulses depends on the conditions of cooling the active material and the pumping tube. With air cooling, the pulse repetition frequency of a ruby laser having an output power of 1 kW is 10 Hz; for a laser with an output of 20 kW the respective figure will not exceed 1 pulse

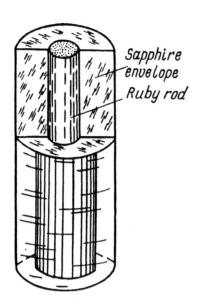

Fig. 24. Ruby rod in a sapphire envelope. The sapphire envelope as if increases the ruby rod dimensions

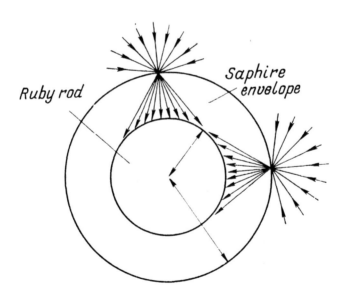

Fig. 25. Scheme of a composite ruby rod illustrating the principle of concentrating radiation on a ruby rod by means of a sapphire envelope

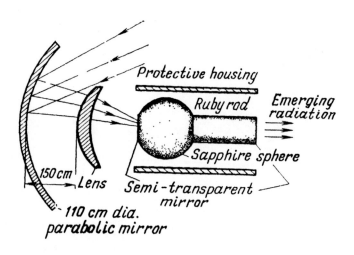

Fig. 26. Solar pumping of a ruby laser

per second. If the system is cooled with liquid nitrogen, the repetition frequency of laser pulses is 30 Hz for the output power of 10 kW and pulse duration of $1\mu s$.

The heating of the active element and flash tube can be diminished by using glass reflectors with an interference coating, which reflects light only in the absorption band of the ruby and transmits the rest of the pumping tube radiation.

Besides optical, there are other methods of excitation, for example, by using solar energy, the energy of exploding wires, cathodoluminescent excitation, etc.

Figure 26 is a schematic presentation of an exciting system using solar energy. As these lines are being written, first experimental sunlight pumping systems are under development. Sunlight is focused by means of parabolic mirrors up to 110 cm in diameter. The light reflected from a parabolic mirror is concentrated by a glass lens onto a sapphire sphere which adjoins

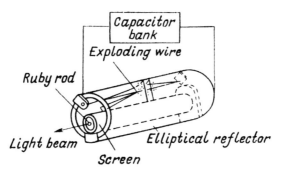

Fig. 27. Exploding wires pumping

one of the end faces of a ruby rod. Instead of a sapphire sphere it is possible to use a glass lens filled with water, which will function both as a focusing and a cooling element. Radiation emerges from the other, semitransparent end face of the ruby.

Another method of excitation is based on using the energy of an exploding wire (Fig. 27). An exploding wire is a very powerful source of light. This wire is placed in one of the focus lines of an elliptical reflector made from stainless steel. A ruby rod is disposed in the other focus line of the reflector. The wire is exploded by a high current pulse.

Experiments with such pumping sources were conducted using tungsten, aluminium and copper wires which were 50 mm long and had a diameter from 0.1 to 0.2 mm. The radiation of a flash created by an exploding wire is most intensive in the ultra-violet region of the spectrum. In the visible region of the spectrum the intensity of such radiation is one third that of a xenon flash tube.

The surface of the ruby rod is protected against drops of molten metal and shock wave effects by a special screen made of glass or plastic.

In improved exploding wire pumping systems an

exploding wire is inserted into a cylindrical rotor and at a certain moment of time this wire is exploded by a strong pulse of current, with the rotor rotating at a very high speed.

This method is most applicable in quantum generators which work pulsed and is rather promising for the creation of superpowerful lasers.

Cathodoluminescent excitation is based on the use of cathode-ray tubes (Fig. 28). The cylindrical surface of these tubes is coated with a special luminescent material called phosphor. A ruby rod is placed axially in the cathode-ray tube cylinder. The phosphor is luminescent when being bombarded with electrons. The radiant energy of such luminescence causes excitation of the ruby rod.

A very interesting and, evidently, promising is a method of exciting one laser by the beam of another laser.

Particularly effective may prove to be a method of pumping a ruby rod with incoherent light of a semiconductor laser (Fig. 29).

Experiments were reported* on pumping active materials by means of nuclear sources. With this method the radiation obtained will be close to X-ray radiation.

RESONANT SYSTEMS

The interaction between electromagnetic radiation and an active material is most effective when this material is placed into a resonant cavity. A resonant cavity usually employed in lasers is a Fabry-Perot interferometer. The cavity of a Fabry-Perot interferometer is an air space bounded by two plane-parallel glass plates facing each other and having thoroughly

---

* Electronic Design, 10, No. 25, p. 96, 1962.

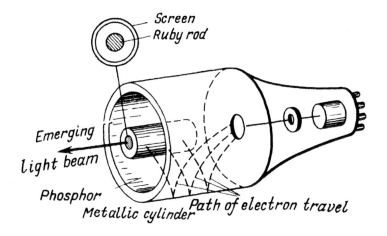

Fig. 28. Cathodoluminescent pumping

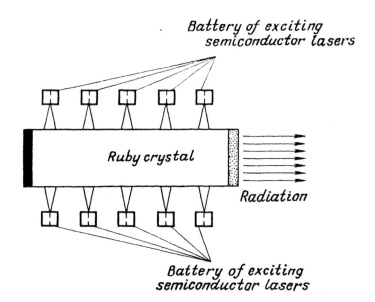

Fig. 29. Pumping of a ruby laser by radiation of semiconductor lasers

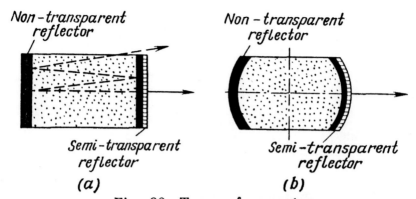

Fig. 30. Types of resonators
(a) resonant cavity with plane-parallel mirrors; (b) resonant cavity with spherical mirrors

ground and polished surfaces that are silvered or otherwise metallised to ensure a high reflection coefficient (Fig. 30a). A light beam, after having entered the interferometer, undergoes a multiple reflection and gives an interference pattern at the exit.

Fabry-Perot interferometers are used for studying fine structures of the spectra. A modification of the Fabry-Perot cavity used in lasers resides solely in that it is not an air layer but an active material which is confined between the reflecting plates.

In a laser with a resonant system the wave travelling from one mirror to the other is amplified while passing through the active medium between them. When the wave reaches the second mirror it loses some of its energy at reflection on account of a finite transmittance of the mirror, and some other part of the wave energy is lost at the edges of the mirrors. For the generation to take place, it is necessary that the total losses of energy conditioned by scattering within the medium, by diffraction and reflection should be less than the energy gained by the light during its passage through the active medium. The mirrors need not be only plane-parallel; confocal spherical mirrors can be used as well (Fig. 30 b).

Systems of resonant cavities with spherical or parabolic mirrors are to meet much less stringent requirements on the accuracy of manufacture and adjustment of the mirrors than plane-parallel resonators, this being a definite advantage. The parameters of the resonant cavity vary but little with small deviations in the radius of curvature of the mirrors or in the distance between them. In resonant cavities losses prove to be most considerable in the optical and infra-red regions on account of absorption at reflection. Therefore their reflecting surfaces are to meet quite stringent requirements.

As has been mentioned earlier, if the active material in a laser is a crystalline substance, the appropriately finished end faces of the rod function as the reflecting surfaces. These end faces are coated either with a metal layer or with several layers of a dielectric film. One of the mirrors is made slightly transparent for the emission to emerge. Sometimes a small aperture is left in the mirror coating.

The reflecting ability of silver is inferior to that of a dielectric coating; this leads to considerable energy losses, especially when silver coatings are applied on the end faces of a ruby crystal which is to operate at high powers.

Besides, silver coatings are liable to deteriorate in service and have to be replaced, since otherwise the output power of the quantum generator will gradually drop and the energy required for the stimulated emission will have to be increased. The properties of dielectric coatings, on the contrary, undergo no adverse changes of such kind and though their application on the crystal surfaces is a more labour-consuming process than silvering, in modern practice preference is given to dielectric coatings.

Sometimes, for ensuring a total internal reflection,

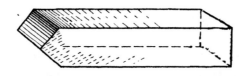

Fig. 31. Ruby rod with one of its end faces shaped as a prism

one of the mirrors in the resonant cavity is substituted by a roof prism (Fig. 31).

In certain cases the mirrors in resonant cavities, though they adjoin the active substance, are made detached. The end faces of such mirrors are ground and polished, but no silver or dielectric coatings are applied on them. If required, the mirror in such a system can be rapidly replaced without disturbing the active substance. Moreover, the mirrors in this system can be mounted parallel to each other with a high precision by means of an adjusting device.

CONTINUOUS-WAVE LASERS

So far we have considered crystal lasers operating in a pulse mode. In some cases it is preferable to have continuously-operating or continuous-wave lasers. CW operation can be obtained with gas and semiconductor lasers which will be discussed in greater detail later. Now we shall only mention that CW gas lasers have a small output power and the emission of semiconductor lasers, likewise capable of continuous operation, is insufficiently monochromatic.

One of the reasons why operation of crystal lasers involves difficulties is as follows. A CW laser requires continuous removal of heat, since otherwise the crystal and the excitation source may become overheated. As is known, only a small part of the radiant energy emitted by the pumping source is actually utilised for the excitation of the active material while a considerable part of this energy is spent on heating of the apparatus.

Special measures are to be taken for a continuous removal of this useless heat so that the laser could operate under the required conditions. The heat removal is effected by a cooling system particularly envisaged for the purpose. However, in case of high pumping powers, a certain period of time is still needed after each excitation cycle for the arrangement to return to its initial temperature conditions and for a balance to be established between the quantity of heat received by and removed from it.

At present several designs of continuously-operating crystal lasers are known. In one of them a sapphire-and-ruby crystal joined for CW operation is employed. This composite crystal consists of a ruby rod and a trumpet-like colourless sapphire light pipe. The sapphire light pipe adjoins with its narrow end face one of the end faces of the ruby rod and serves for gathering the incident exciting light produced by the pumping source which is a mercury-vapour arc lamp. The radiation spectrum of this lamp abounds in green and violet rays prerequisite for the excitation. The emission of the mercury lamp is focused on the trumpet-like part of the sapphire pipe by a spherical mirror. The ruby rod is cooled with liquid nitrogen. The size of the crystal is very small: its length is 1.15 cm and its diameter is 0.061 cm. The length of the sapphire pipe is 1.05 cm and the diameter of its trumpet is 0.15 cm. The surface of the ruby rod is optically polished. The tolerances within which its end faces are kept plane-parallel are 0.1 of the laser operating wavelength. With the pumping power of 930 W the power of the CW-laser emission is 4 mW.

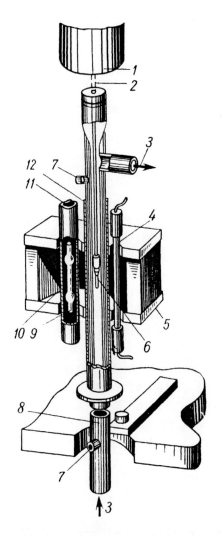

Fig. 32. Design of a continuously-operating laser

1—radiation detector; *2*—laser beam; *3*—liquid oxygen; *4*—xenon flash tube; *5*—water cooling of elliptical reflector; *6*—crystal of active material; *7*—liquid filter; *8*—dewar; *9*—mercury lamp; *10*—elliptical reflector; *11*—water jacket; *12*—heat insulation

In another form of a continuously-operating laser (Fig. 32) the crystal is calcium tungstate activated with trivalent neodymium ions ($CaWO_4:Nd^{3+}$). This material has a relatively low pumping threshold. The crystal shaped as a rod with confocal spherical end faces is placed in one of the focus lines of an elliptical cylinder and surrounded by a liquid filter which does not pass the ultra-violet and infra-red rays of the light emitted by a pumping mercury tube positioned in the other focus line. In this laser the crystal is cooled with liquid oxygen; the excitation source and the reflector are provided with a water cooling system to protect the apparatus against overheating. The mercury tube consumes about 900 W and the laser output power ranges within 3 to 5 mW, its operating wavelength being 1.065 µ. For the laser to be capable of pulsed working, the reflector, besides the mercury tube, also houses a xenon flash tube.

Other lasers of a similar design have been constructed, in which the crystals are 38 mm long and 3 mm in diameter. With the pumping power of the order of 600 W their output power ranges from 0.5 to 1 W.

For increasing the output power of lasers, systems with more powerful excitation sources are suggested. In one of them the excitation source is a 500 kW arc furnace filled with a gas under pressure. The radiation emitted by the electric furnace is focused on the crystal surface by two spherical mirrors. The ruby crystal is housed in a spherical cavity made of pyrex and filled with a solution of metallic copper which absorbs infra-red, red and ultra-violet rays and transmits light within the range of 4500 to 6000 Å. This solution also serves for cooling the ruby crystal. A laser of this construction could operate continuously for a period of 0.2 to 1.2 s. A more prolonged excitation caused overheating and destruction of the crystal.

Besides the above-considered crystals for use in CW lasers, about ten other crystals most fit for the purpose are known at present (see Table 2).

GLASS LASERS

The development of lasers with glass as the active material doped with rare-earth elements such as neodymium, ytterbium, gadolinium, holmium or terbium is of great interest. The design of glass lasers is essentially the same as of the crystal ones. An important advantage of glass lasers resides in that glass rods of any required size and shape are relatively easy to manufacture, this being a prerequisite for the creation of lasers with a high output power. Glass batches can be pulled to fibres for making optical waveguides.

The distribution of the energy levels of rare-earth ions in glass is almost the same as in crystal matrices. But, unlike crystals, glass features no definitely oriented and regular structure; therefore spectral lines of emission in it are somewhat broader than in crystals.

One of the first glass lasers, which have become widespread nowadays, was a laser with neodymium-doped barium glass as the active medium. The concentration of neodymium in glass may vary from 0.13 to 10 per cent. The laser behaves as a four-level system with the induced radiation taking place at the wavelength of 1.06 µ.

Early glass lasers employed thin barium glass rods 76 mm in length coated with a glass layer that had a somewhat smaller refractive index than the rod material. This increased the translucence of the rod and ensured a more effective excitation. Yet, on account of a strong absorption displayed by neodymium ions, the penetration of the pumping energy into the glass was poor and the quantity of this energy was not sufficient for exciting the inner layers of the glass rod. The rods, therefore, had to be of a small diameter (0.3 and 0.032 mm in the test specimens). The parallel end faces of the rods were polished and coated with a layer of silver having a 2% coefficient of transparency. The laser worked pulsed at room temperature; pumping was effected by means of a xenon flash tube. An essential disadvantage of the first lasers with thin glass rods as the active material was their low output power.

By now several models of glass lasers with a high stimulated emission energy have been developed. Thus, glass lasers have been created with rods having a length of about half a metre and a diameter of several centimetres. Their output power is over 100 J.

If a glass laser beam is allowed to pass through a

Fig. 33. Soviet neodymium glass laser ГСИ-1

crystal having definite non-linear optical characteristics, a green beam will emerge from this crystal as a result of the appearance of radiation harmonics.

Glass lasers where ions of rare-earth elements other than neodymium are used as dopants differ from the laser discussed above by the wavelength of the induced radiation. Thus, the operating wavelength of glass lasers doped with ytterbium is 1.015 μ; of those with holmium, about 1.95 μ; with gadolinium, 0.3125 μ; and with terbium, 0.535 to 0.55 μ.

Figure 33 is a photograph of the Soviet ГСИ-1 glass laser designed for studying the interaction of light with the substance. Neodymium glass plates $8 \times 45 \times 150$ mm in size are used as the active material in this laser. Pumping is effected by eight flash tubes.

The laser generates light pulses of 0.7 ms duration at the wavelength of 1.06 μ; its maximum energy is 75 J. The weight of this laser together with its power supply unit is 200 kg.

GIANT PULSES

In the creation of lasers, alongside of the main trending towards increasing their output energy, efforts are made to increase the power of individual pulses.

As has been pointed out in our previous discussion, a laser pulse, if "expanded" along the time scale, will display a number of spikes or peaks. The laser energy is radiated during a comparatively long period of time which is determined by the pumping time and usually amounts to several milliseconds.

However, there is a possibility of concentrating the radiation within a time interval as short as milliardth fractions of a second. The power of such a pulse, with the radiation energy remaining the same, increases by millions of times (it will be recalled that power is equal to energy divided by time). Tremendous amounts of power concentrated in extremely short pulses received the name of "giant pulses". The power of these pulses reaches dozens of megawatts and even more.

What are the methods of producing giant pulses?

As is known, in conventional lasers the radiation starts while pumping is still going on, and this circumstance is a hindrance to bringing the majority of active particles to an excited state. In other words, no great difference in the inverted population of energy levels can be attained.

Nevertheless, there can be obtained such a state of the active medium when almost all its active particles will be excited. This state can take place before the onset of generation. Consequently, for these conditions to be ensured, the generation should be precluded till the pumping process has been accomplished. As we know, generation in a laser is ensured by the provision of a resonant cavity in it. If during pumping the resonant cavity in the laser system is kept inoperative, no generation will take place and, as a result, most of the particles in the active medium will be brought over to the excited state.

One of the methods of obtaining giant pulses in operating laser systems consists in placing an optical shutter between the laser rod and one of the mirrors. At the moment of pumping, when the flash tube is switched on, the shutter does not pass light to the mirror and no reflection occurs. Though a very great number of particles of the active material are brought to the excited state, generation does not take place. When pumping is over and the number of excited ions reaches a maximum, the optical shutter is rapidly opened. As soon as it is done, generation is immediately set up in the rod. All the energy accumulated in the excited ions is released in a single giant burst pulse which lasts for about 10^{-9} s and has a power of the order of 50 MW.

The principle by which giant pulses are obtained can be understood from a consideration of Fig. 34.

One of the most commonly employed optical shutters is a Kerr cell.

As has been mentioned before, the radiation of a ruby crystal, in case its optic axis does not coincide with the axis of the ruby rod, is polarised. A Kerr cell filled with nitrobenzene is arranged in such a manner that the vector of the electric field applied to it should be at 45° to the plane passing through the optic axis and the axis of the ruby rod. The plane of polarisation of the light which traverses the cell twice (towards the mirror and back) is rotated through 90° in relation to the initial direction. The reflected light proves to be insufficient for generation in the laser rod.

For obtaining a powerful pulse, first the flash tube is switched on, and then, in 0.5 ms, the Kerr cell is de-energized during 0.02 ms. The optical shutter being

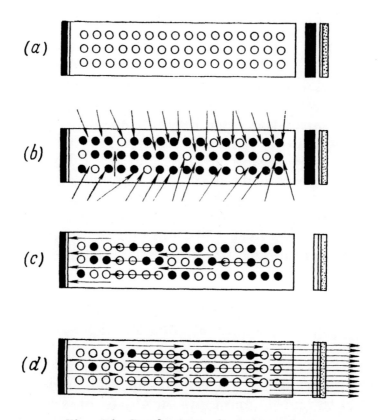

Fig. 34. Producing of giant pulses

(a) crystal of active material; shutter is placed between the crystal and one of the mirrors; (b) the crystal is pumped, active particles are excited, no generation takes place; (c), (d) the shutter is open, generation takes place, and energy is released in one giant pulse

thus opened, the generation follows immediately and proceeds with a higher degree of excitation than in a conventional laser having no optical shutter. A giant light pulse emerges from the laser.

Instead of a Kerr cell, an ultrasonic cell filled with alcohol or kerosene can also be used as an optical shutter. In this case the structure of the substance which fills the cell undergoes changes on account of transverse compressions and rarefactions caused by ultrasonic vibrations usually produced with the help of a high-frequency oscillator. These changes in the structure of the cell determine its refractive index and the degree of light scattering. When the light scattering is zero, i.e. when the substance in the ultrasonic cell is most transparent, conditions necessary for the generation are created. During one period of vibration such favourable conditions occur twice and therefore the repetition frequency of laser pulses is twice that of ultrasonic vibrations in the cell.

Rotary disks and mirrors are also employed as optical shutters. With these shutters it is possible to produce both individual pulses and series of pulses.

Though all the above-described methods have found practical application, yet they prove to be rather complicated and require additional equipment. Therefore, still another method is used for producing giant pulses, in which so-called brightening filters are employed. The brightening filter in this case is a solution of phthalocyanine. Under the action of light this liquid changes its colour and transparency. A simple small cuvette with a phthalocyanine solution is placed between the laser rod and one of its mirrors. The solution strongly absorbs light at the frequency of the ruby generation and therefore the resonant cavities of the laser do not amplify the light until a considerable number of chromium ions in the ruby crystal have been pumped into the upper energy state.

When the pumping energy reaches a value at which the ruby amplification exceeds the absorption losses in the phthalocyanine solution, the laser will start rather weakly radiating coherent light. Nevertheless, a small quantity of this additional light proves sufficient for the solution to become colourless to such an extent that all of a sudden it turns out to be absolutely transparent. At this very moment the generation will be sharply increased and all the energy accumulated in the ruby will be instantaneously emitted as a giant pulse. The pulse having been emitted, the solution regains its absorbing capacity and the next pulse can be formed in a similar manner.

In an experiment conducted for obtaining giant pulses the cuvette with a solution of phthalocyanine about 30 mm long was placed between a ruby crystal and a mirror having a rather high coefficient of reflection. The other end face of the ruby was shaped as a 90° prism for ensuring internal reflection and necessary generation conditions (see Fig. 31). The concentration of the solution was selected such as to provide a 50 per cent transmission of light at low energies. When the ruby rod was pumped to an energy at which under usual conditions (with no solution) the onset of generation took place, a giant pulse lasting for about 20 nanoseconds was observed. At higher pumping energies a series of giant pulses was observed, with intervals between the pulses of the order of 100 μs. The power obtained in this experiment was 50 MW.

It is believed that such a method can ensure obtaining of much higher power values.

There are methods by which a giant pulse can be amplified again, so that a still more powerful pulse will be produced. To this end, a second rod from an active material is arranged in the path of a giant pulse. This can be, for example, a ruby rod with unsilvered end faces. In this case the second rod functions merely as an amplifier and does not generate. Such an amplifier must be pumped by means of a special flash tube. The light of the latter brings almost all the particles of the active material of the rod to the excited state. The second tube is switched on somewhat later than the first. When passing through the second rod, the giant pulse causes induced radiation in it, which is added together with the giant pulse. As a result, the emerging giant pulse is more powerful. Since induced transitions are caused already by the leading edge of the pulse entering the rod, the newly formed pulse becomes more intensive, yet of a shorter duration. The power of the pulse in this case may reach 500 MW.

GAS LASERS

Concurrently with crystal lasers wide application has been found by lasers in which gases and vapours of metals are employed as active media. Such lasers are usually called gas lasers and their main advantage resides in that they operate continuously, though some gas lasers are also capable of pulsed working. Gas lasers display exceptionally high monochromaticity, most pure spectrum and high stability of frequency. All these features make gas lasers extremely useful in various branches of science and engineering; probably, the most wide application of these lasers will be in communications. The output power of gas lasers, however, still remains rather moderate and much inferior to that of crystal lasers.

The first gas laser developed by A. Javan, W. Bennett and D. Herriot operated on the principle of resonant transmission of excitation energy in a gas discharge. The gas laser is a fused quartz tube with a diameter of about 1.5 cm and 80 cm long. This tube is filled with a mixture of gases: neon (Ne) under a pressure of 0.1 mm Hg and helium (He) under a pressure of 1 mm Hg. The tube is connected with flexible glass-to-metal seals with metallic heads ac-

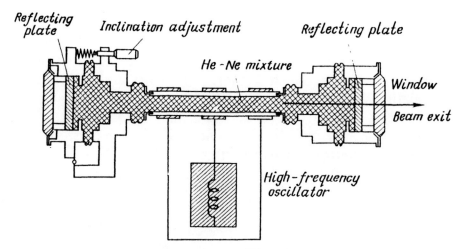

Fig. 35. Design of a helium-neon laser

commodating plane reflecting plates. A simplified diagram of such laser is shown in Fig. 35.

Insofar as during its passage through the active He-Ne mixture the light beam is amplified but to a small extent (of the order of 2 per cent per metre of the tube length), much attention is paid to the quality of the reflecting plates. Their surfaces are ground and polished to a tolerance in flatness of 0.01 of the laser wavelength. For ensuring better reflection conditions multilayer dielectric films are also employed. A film consisting of 13 alternating zinc sulphide and magnesium fluoride layers allows the obtaining for waves within the region of 11 000 to 12 000 Å the coefficient of reflection of 98.9 per cent, coefficient of transmission of 0.3 per cent, the absorption and scattering losses being 0.8 per cent.

All the elements of the active part of the laser are annealed and degased under ultra-high vacuum conditions. Should a single drop of water condense on the dielectric reflecting layers when the evacuation process is started or atmospheric air be present in the system after the annealing, this will damage the layers. The position of the reflecting plates is adjusted with the help of a micrometer device employing flexible bellows. The initial adjustment of the plates is performed before starting the generator.

A system of mirrors makes up a Fabry-Perot resonant cavity which ensures an optical feedback required for the creation of self-excitation conditions and generation.

The active mixture is excited by means of a high-frequency generator with a frequency of several tens of MHz and an input of about 50 W. The electromagnetic field in the gas mixture is set up by means of external electrodes that encompass the quartz tube of the laser. The mechanism of excitation of atoms in gas lasers is quite different from that in crystal lasers and therefore requires a more detailed explanation.

The operation of a gas laser is based on the interaction of atoms of two gases that are in close energy levels (Fig. 36). As can be seen from the energy diagram shown in Fig. 36, the level 2^3s of helium lies close to the level $2s$ of neon, which consists of four sub-levels. The atoms of helium are excited by the gas discharge and pass to the upper level 2^3s. On account of inelastic collisions between the atoms of the two gases, the excited atoms of helium give off their energy to the atoms of neon, so that the latter rise to one of the four metastable levels $2s$.

When the population of the neon level $2s$ becomes sufficient, an induced coherent radiation corresponding to the transition to the level $2p$ sets on. The neon atoms then return to their ground state. The neon level $2p$ consists of 10 sub-levels. The total number of possible transitions corresponding to diffe-

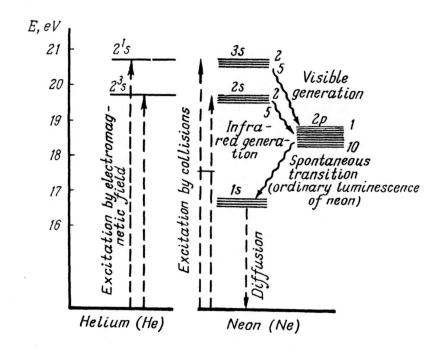

Fig. 36. Energy levels of helium and neon

rent lengths of the radiated waves is 16, all of them lying in the near infra-red region within 9400 to 15 500 Å. By now generation has been obtained on five waves, the best results in terms of intensity being for the wavelength of 11 533 Å. Recently generation in the visible region at 6328 Å was produced.

The emission of a gas laser is highly monochromatic and coherent. The angular divergence of a gas laser beam is less than one minute. The width of the spectral line of a gas laser emission is approximately one hundred thousandth that of a ruby laser. In the general case a gas laser linewidth lies within 10 to 80 kHz. The elimination of so-called microphonic effect (jitter of mirrors) has lately allowed obtaining of a He-Ne gas laser linewidth of about 1 kHz.

Fig. 37. A helium-neon laser

With the generator input power of about 40 to 90 W, the laser output is 0.5 to 10 mW.

One of the designs of a He-Ne laser is shown in Fig. 37.

Experimental models, in case optimal generation conditions are ensured as to the pressure and discharge current values, can have an output of up to 100 mW. The length of a helium-neon laser displaying such a power is about 2 m, and the diameter of its tube is 7 to 10 mm.

Though, for increasing the radiation power, such a laser can be made with a longer tube, this is not always advisable, since practical use of the apparatus will be more difficult. Any further increase in the tube diameter will give no positive effect: on the contrary, the output power of the helium-neon laser will be diminished.

The helium-neon mixture is not the only active medium which can be employed in gas lasers: a mixture of argon and oxygen (Ar-O_2) can also be used for this purpose. The operation of lasers with such an active medium is based on the excitation of dissociated molecules of oxygen. An atom of the inert gas (argon) excited by a high-frequency discharge, when colliding with a molecule of biatomic oxygen, becomes excited to a higher level, this resulting in an inverted population, while another oxygen atom absorbs the remaining energy. Emission takes place at 8445 Å. The output power of Ar-O_2 lasers is about 2 mW. The same principle is employed in a laser with a mixture of helium and carbon oxide (He-CO) as the active medium.

Single-gas lasers have also developed; they use inert gases—helium, neon, argon, krypton, and xenon—taken individually as the active medium. All these lasers generate in the near infra-red region of the spectrum at several tens of different wavelengths. Maximum power is obtained with a xenon laser (about 5 mW). The operation of this kind of gas lasers is based on direct excitation of atomic levels due to inelastic collisions with electrons in a gas discharge.

Figure 38 is a photograph of a Soviet gas CW laser. Its output power is 1 mW and it is designed for conducting various investigations in physics, chemistry, medicine and radiobiology.

At present a CW output of 11.9 W is obtained with transitions corresponding to the wavelength of 10690Å. Power radiated at the most intensive frequency is about 75 per cent of the total output. The laser-conversion efficiency is about 3 per cent. The experimental laser system employed mirrors placed inside the laser tube provided with a special device which allowed mixing the gaseous substances in a continuously circulating flow. The discharge was obtained by passing direct current through the active zone of the laser, having a diameter of 25 mm and a length of 2 m. The optical resonator consisted of a concave mirror with a radius of curvature of 11 m and a convex mirror with a radius of curvature of 10 m. The

Fig. 38. One of Soviet continuously-operating gas lasers, Model ОКГ-11

mirrors were spaced 240 cm apart. The energy from the resonator emerged through a 8 mm circular aperture made in the centre of the concave mirror. The both mirrors were coated with gold.

The above-described laser is the first apparatus capable of a high-power emission in the infra-red region at the wavelength of 10690Å. It is expected that with greater dimensions of the laser tube a considerable gain in the power of coherent radiation can be obtained in CW operation.

This laser is also advantageous in that its radiation stability is almost unaffected by external electric and magnetic fields. The high power of coherent radiation of the gas laser will make possible the study of non-linear optical phenomena in the infra-red region of the spectrum. This laser proves to be of great interest for communication engineering as well, since its radiation wavelength lies within the range of 8 to 14 μ where the atmospheric absorption of light is insignificant.

One of the latest developments is a whole group of gas lasers in which the active medium are strongly ionized inert gases, as well as sulphur, chlorine and phosphorous vapours. Ionisation in such lasers is usually effected by means of an arc discharge with a very high current density (several thousand amperes per square centimetre). These are so-called ion gas lasers or, simply, ion lasers. Most ion lasers work pulsed, though some of them, under certain conditions, are capable of continuous operation.

In such lasers generation is not due to atomic transitions, as in the helium-neon laser, but to transitions between excited energy states of an ionised gas.

Ion lasers are most powerful sources of coherent radiation in the visible and ultra-violet regions. The efficiency of these lasers remains not high and reaches only 0.01 to 0.1 per cent.

Shown in Fig. 39 is a powerful pulsed gas laser ЛГИ-37 made in the Soviet Union. This laser radiates in the visible region of the optical spectrum. It is excited by voltage pulses of 20 to 25 kW. The power radiated in a pulse is about 2 kW. The gas discharge tube of the laser is water-cooled.

Most typical in the group of ion lasers is an argon laser. Inversion of the population in ionised argon occurs (see Fig. 40) when its ions are directly excited from the state *1(3p^5)* to the state *3(3p^{4}4p)*. Some transitions to the excitation stage proceed stepwise. Inverted population in the *3-2* transition takes place mainly on account of a short lifetime of argon ions in the state *2*, which is approximately 4 per cent their lifetime in the state *3*. The level *2* is depleted rather quickly, so that the inversion of the population and generation can be maintained continually.

Fig. 39. Powerful pulsed-working gas laser ЛГИ-37

The levels *3* and *2* consist of groups of sub-levels and therefore generation can take place at several frequencies simultaneously. The output radiation of an argon ion laser is predominantly green and blue (5145 Å and 4880 Å wavelengths, respectively).

Typical output power figures are unities of watts, but some models can give up to 150 W. Theoretical calculations show that in the not distant future the power of argon lasers can be increased to several hundred watts.

Structurally an ion argon laser is a narrow water-cooled fused silica capillary tube in which an arc discharge takes place (see Fig. 41). Arranged at the ends of this capillary tube are an anode and a cathode. The anode and the cathode spaces communicate through a by-pass gas tube which ensures free circulation of the gas. For increasing the laser output power and efficiency, sometimes a solenoid or a permanent magnet is put on the capillary tube, its fun-

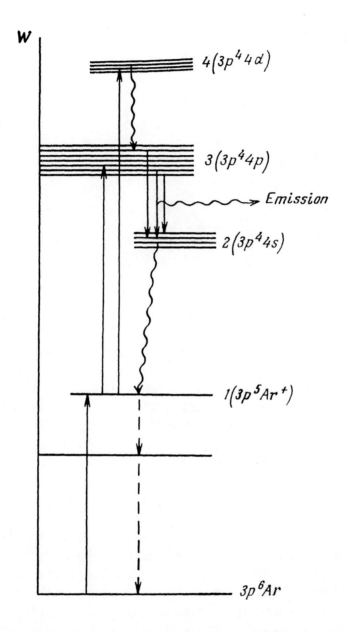

Fig. 40. Diagram of energy levels of ionised argon Ar⁺

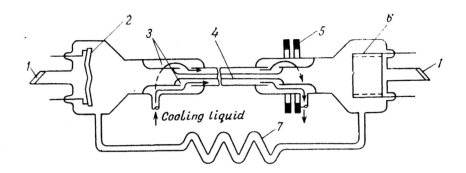

Fig. 41. Design of an argon laser

1—outlet windows; *2*—cathode; *3*—water cooling channel; *4*—arc discharge channel (capillary); *5*—magnet; *6*—anode; *7*—by-pass gas tube

ction being to constrict the discharge area and increase the concentration of ions along the capillary axis. Fused silica capillary tubes, however, cannot ensure long-term service of the laser. Therefore, for extending its service life, metallic capillary tubes are resorted to, made as cylinders partitioned by thin ceramic rings.

Shown in Fig. 42 is a Soviet argon laser "Malakhit". This laser is designed for conducting laboratory investigations in physical optics, spectrometry, telephone and television communications, and holography. The laser radiates in the indigo-green region of the spectrum at 10 wavelengths: 0.4545; 0.4579; 0.4609; 0.4658; 0.4726; 0.4765; 0.4880; 0.4965; 0.5017 and 0.5145 μ. Its output power is 0.2 to 1.0 W, input power is 5 kW. The laser is water-cooled and weighs 40 kg.

The main disadvantage of gas lasers employing atomic and ion transitions is their low efficiency, which can be accounted for by a small effectiveness of the electron pumping mechanism. Gas lasers in which the mechanism of the oscillatory states of

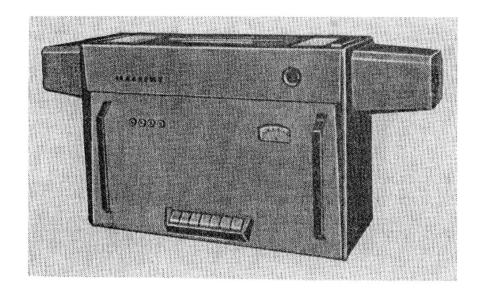

Fig. 42. Argon laser "Malakhit"

molecules is employed proved to be rather promising in this respect. These are so-called molecular gas lasers.

The energy state of a molecule is known to be determined not only by the internal energy of the atoms of which it consists, but also by their oscillatory motions in the molecule. These states are quantised, they have definite allowed levels and are characterised by a certain set of discrete values. In the molecule, as in the atom, there are non-excited energy levels with a minimum value and excited states with a higher energy. Naturally, the more complicated the molecule, the more complicated the structure of its oscillatory states is.

At present in gas lasers employing molecular energy transitions generation is effected in carbon monoxide, carbon dioxide, molecular nitrogen and nitrogen monoxide.

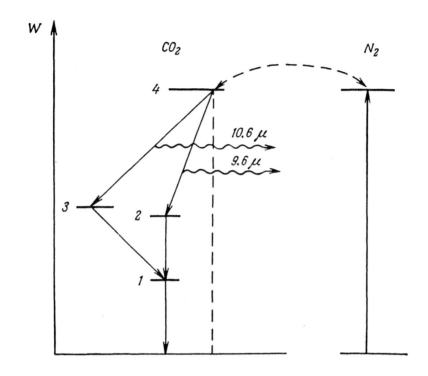

Fig. 43. Diagram of oscillatory levels of CO_2 and N_2 molecules

Let us consider a gas laser with carbon dioxide as the active medium. The working material in such a laser is a mixture into which, besides carbon dioxide, nitrogen and, sometimes, for increasing the radiation power, also helium or water vapours are added.

Fig. 43 shows energy diagrams of oscillatory levels with which the main physical processes taking place in this laser are associated.

The molecules of carbon dioxide are excited by resonant energy transfer effected by excited molecules of argon. Generation sets on at transitions *4-3* and *4-2*. Radiative transition *4-3* corresponds to the wavelength of 10.6 μ and transition *4-2*, to 9.6 μ.

The design of a CO_2 laser is shown in Fig. 44.

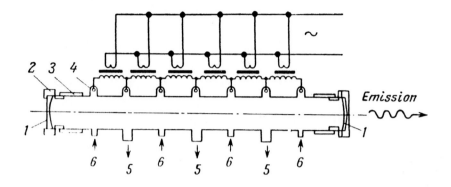

Fig. 44. Design of a powerful CO_2 laser

1—mirror; *2*—mirror mount; *3*—flexible connection; *4*—electrodes; *5* and *6*—gas mixture evacuation and supply

A positive feature of this laser is the dependence of the radiation power on the tube diameter, therefore the output power can be raised by increasing the tube diameter. In powerful carbon dioxide lasers the length of the gas discharge tube (or cuvette) may be several metres and its diameter, several centimetres. For the glow discharge to be sustained in such a cuvette, the latter is divided into several individual sections. The laser is powered with a.c. or d.c. of an industrial frequency (50 Hz). A serious problem with powerful CO_2 lasers is the manufacture of durable reflecting coatings (mirrors). For low-power lasers the mirror surface is manufactured from a multi-layer dielectric. In powerful lasers metallic mirrors are employed, mostly from gold. Since such a mirror cannot be made semitransparent, a small aperture is left in it for the exit of radiation. These lasers are either water-cooled, or employ forced-air cooling systems. Their efficiency is 20 to 30 per cent.

At present the method of raising the laser output power by pumping gas through the discharge tube becomes most widespread. This procedure is necessary insofar as the gas mixture decomposes in the course of the laser operation. The gas can be pumped through either longitudinally or transversely of the gas discharge tube. The laser radiation power can be strongly increased if the gas is pumped through at such a rate that it would have no time for being heated to the critical temperature under the effect of the discharge current. Thus in a tube having a length of 10 cm and a diameter of 1.35 cm with the longitudinal pumping-through rate of about 100 m/s, the power obtained under continuous operation conditions was 140 W. In case a longitudinal pumping-through is used, the gas rate should increase with an increase in the length of the gas discharge tube. Therefore lasers with the transverse pumping-through are considered to be most promising and convenient. Transverse pumping-through calls for modifications in the conventional design of a laser. The walls in such a laser are two plane plates. The discharge glows and the laser radiation propagates in one direction, along the plates, and the gas is pumped through in the other direction, perpendicular to the first. The necessary effect is attained at lower rates of pumping-through than in lasers where the gas is pumped through longitudinally of the gas discharge tube.

Already in the first models of such lasers with the resonant tube 1 m in length, with the transverse pumping-through of the gas and preliminary cooling of the mixture (to diminish the population of the lower energy level), the output radiation power attainable was about 1000 W. An increase in the length of the gas discharge tube leads to an almost proportionate increase in the output power. The output power attained recently in CW operation is close to 60 kW.

The significance of such power will be perhaps better appreciated if we say that a 100 W laser beam can burn a brick through. A 60 kW beam is capable of destroying rocks. Such a laser can be employed for tunnelling, for mining, for power transmission, etc., the more so that the power of 60 kW so far attained is not at all the limit. Table 3 gives the main characteristics of active media for gas lasers.

Table 3
Main Characteristics of Active Media for Gas Lasers

Active Medium	Radiation wavelength, Å	Laser operation
Ionised neon, Ne^{3+}	2 358	Pulsed
Ionised neon, Ne^{2+}	3 324	Pulsed
Molecular nitrogen, N_2	3 371	Pulsed
Ionised argon, Ar^{2+}	4 880	CW
	5 145	CW
Ionised krypton, Kr^{2+}	5 682	CW
Helium and neon mixture, He+Ne	6 328	CW
	11 523	CW
	33 920	CW
Argon and oxygen mixture, $Ar+O_2$	8 445	Pulsed, CW
Mixture of helium and cadmium vapours, He+Cd	3 250	CW
	4 420	CW
Helium and carbon oxide mixture, He+CO	14 540	Pulsed, CW
Atomic xenon, Xe	20 610	Pulsed
Caesium vapours, Cs	71 821	CW
Mixture of carbon dioxide, nitrogen and helium, CO_2+N_2+He	106 000	CW, pulsed
Water vapours, H_2O	279 000	Pulsed
	1 186 000	Pulsed
Hydrocyanic acid, HCN	3 370 000	CW

METHODS OF CONCENTRATING GAS LASER RADIATION AT ONE FREQUENCY

In our discussion of the laser radiation we more than once emphasized its being highly monochromatic. Indeed, as compared to the emission of conventional light sources, the monochromaticity of radiation of all kinds of lasers is extremely high. Maximum monochromaticity is displayed by the gas laser: its radiation line width is one hundred thousandth that of a ruby laser.

Now, if we "expand" the scale of frequencies of a gas laser radiation, we shall see that the radiation in question, though it occupies a certain very narrow band, features a number of peaks. The distance between these peaks depends on the distance between the mirrors of the resonant cavity employed. Thus, if the distance between the mirrors is 1 m, the separation between individual frequency peaks will be 150 MHz. The amplitude and phase of each vibration mode do not depend on the amplitude and phase of other vibration modes. The phases of these vibrations are random in character.

In conventional gas lasers the dimensions of their resonant cavity considerably exceed the optical radiation wavelength. With the cavity in the resonant state, an integral number of waves of many different modes can occur between the mirrors. This accounts for the resonance taking place at different frequencies.

Numerous attempts have been made at creating a gas laser which would operate at one frequency. To this end, the following three methods can be resorted to:

1. Diminishing the length of the resonant cavity to such an extent that only one resonant frequency

of the Fabry-Perot resonator should correspond to the amplification range of the active medium.

2. Subjecting the resonant cavity to attenuation, or diminishing the amplification of the active medium in such a manner that the amplification should exceed losses in the resonant cavity only at one vibration mode.

3. Placing the mirrors of Fabry-Perot resonators employed in conventional helium-neon lasers farther apart, so as to ensure greater separation of the frequencies.

These three methods do give the desired result, but at the expense of other performance characteristics of the laser (its output power is reduced or frequency stability lowered).

A new method which was suggested recently allowed the laser radiation to be concentrated at one frequency, without making the system disadvantageous in the above-mentioned respects.

This was attained by incorporating a phase modulator in the Fabry-Perot cavity of a conventional He-Ne laser.

If the modulator operates at the frequency equal to the difference in the frequencies of neighbouring peaks of the resonant cavity oscillations, say of 100 MHz, then the distribution of the amplitudes and phases in the total emission becomes similar to that it would be in the side bands of a frequency-modulated signal. All the radiation energy of the laser is distributed in the side bands, substantially at two frequencies.

Modulation is effected by means of a crystal on which a high-frequency voltage is impressed. The crystal is placed between the laser tube and one of the mirrors, i.e. inside the resonant cavity. The electric field induced in the crystal is oriented parallel

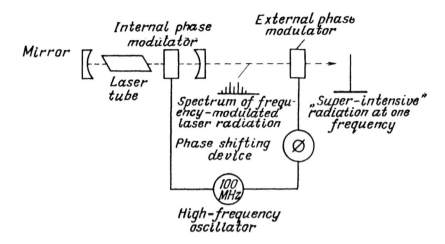

Fig. 45. Basic diagram illustrating the method of concentrating gas laser radiation at one frequency

to the direction of polarisation of the laser radiation. On account of variation of the voltage impressed on the crystal, the length of the optical path in the resonant cavity also varies. This phenomenon resembles the effect observed when one of the mirrors in the resonant cavity is vibrating. The frequency of such vibration corresponds to that of a high-frequency oscillator. The ultimate result is phase modulation.

Another method consists in passing the frequency-modulated laser radiation through a second phase modulator which is phase-shifted by 180° in relation to the first modulator and has the same modulation factor as the first modulator. With this method monochromatic "super-intensive" emission can be produced at one frequency. This emission is noted for the absence of many kinds of noises present in the emission of a conventional laser.

The principal diagram of operation of a single-frequency laser is shown in Fig. 45.

Good spectral characteristics of the above-considered laser system in which both of the last-mentioned methods are employed make it fit for use in communication equipment as a carrier frequency oscillator and as a local oscillator in optical receivers and radars.

LIQUID LASERS

Soon after the creation of lasers in which solid and gaseous substances were used as the active media, it was established that generation could be obtained in certain liquids as well.

We know that the main advantage of solid materials as active media is a high concentration of active particles per unit volume. This makes possible the obtaining of high radiation powers. The dimensions of solid active elements, however, are limited by technological factors: thus, it is difficult to ensure high optical homogeneity of the crystal or glass and high-quality finishing of the rod end faces. When heated, the active solid material is liable to destruction.

Gas lasers are capable of ensuring a high average radiation energy and a small divergence of the beam. But since the density of the active medium in gas lasers is not high insofar as they operate at a very low pressure, attempts at raising their radiative energy lead to a considerable increase of the dimensions of the gas laser. Most powerful gas lasers are up to 10 m in length and 1-2 dm in diameter.

Liquid lasers combine the advantages offered by solid-state and gas lasers. Liquid active media are capable of ensuring a high optical homogeneity in a large volume together with a considerable concentration of active particles in it. The problem of cooling becomes less complicated, since the required temperature can be maintained with the help of an external heat exchanger.

First liquid lasers employed solutions of organic complex compounds (europium chelates). These liquids, however, have an excessively high light-absorption coefficient.

Present-day liquid lasers employ inorganic liquids and organic dyes as the active media. As the inorganic liquid use is made of phosphorus oxychloride or selenium oxychloride with the addition of tin tetrachloride or other metal halides. The activator is a few per cent solution of neodymium oxide Nd_2O_3. Such a laser resembles a glass laser. But in a liquid laser the active medium line width is substantially narrower (of the order of 1 Å) due to the higher homogeneity of the active medium.

Structurally the active element of a liquid laser is a liquid-filled cylindrical cuvette made from high-quality glass or fused silica. The excitation is by optical pumping.

A rather promising method is the use of solutions of organic dyes. Its main advantage is the possibility of generation at any frequencies in the visible and near infra-red regions of the spectrum and of smooth frequency retuning.

The optical pumping source can be a conventional flash tube. A ruby laser or a neodymium glass laser can also be used for optical pumping. In this case an organic dye laser functions as a frequency changer. The resonant cavity, besides the cuvette with an organic dye, also accommodates an interferometer-selector for narrowing the emission spectrum and frequency retuning.

In many cases the laser-excitation efficiency reaches 50 per cent. Among organic dyes generating at visible-spectrum frequencies pyronine, rhodamines, trypaflavine are the most effective. Conventionally employed solvents are alcohol, glycerol, sulphuric acid, water.

Thanks to the high gain factors these organic dyes are suitable for creating wide-band quantum amplifiers of light.

SEMICONDUCTOR LASERS

In crystal, liquid and gas lasers stimulated emission is excited by means of light or a gas discharge. In semiconductor lasers active media are semiconductor materials, and excitation is effected directly by electric current. Semiconductor lasers have a high efficiency which in the existing models reaches 60 to 70 per cent. It is believed that even a 100 per cent efficiency is attainable with semiconductor lasers. However, these are not the only features which make semiconductor lasers worthy of attention. Semiconductor lasers allow easy variation of their radiation frequency with the help of a magnetic field and at the same time are capable of ensuring a high stability of the output frequency, which is characteristic only of gas lasers. Modulation in semiconductor lasers is most simple to effect—by using exciting current.

There can be no doubt that these remarkable properties of semiconductor lasers will guarantee them wide application. It is true that in their monochromaticity, coherence and beam divergence characteristics semiconductor lasers cannot compete with gas and crystal lasers, being rather inferior to them. The output power of semiconductor lasers is not high either. In these lasers the linewidth-to-wavelength ratio is approximately $1 : 10^5$, whereas in gas lasers this figure is $1 : 10^{10}$ and even $1 : 10^{11}$. But one should bear in mind that the semiconductor laser is the youngest in the family.

The appearance of semiconductor lasers has extended the range of materials which can be used in light generating systems, offered new possibilities for producing an active medium.

How does a semiconductor laser operate? In what way is a non-equilibrium energy state required for the amplification and generation of light established in such a laser?

Unlike individual atoms, semiconductors do not have separate energy levels. Semiconductors display groups of energy levels, so-called *bands*, which are arranged in a continuous succession (Fig. 46). The upper group of levels is termed a *conduction band* or an *empty band*, the lower group is called a *valence* or *filled band*, and the separation between these two bands is called the *bandgap* (or forbidden region). If an electron occupying one of the energy levels of the valence band is imparted an additional energy, it will pass to a higher energy level in the conduction band. This will result in the appearance of a positive

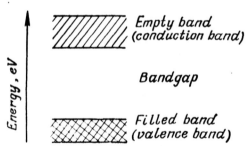

Fig. 46. Energy bands in a semiconductor

charge carrier in the valence band, this being the vacancy the electron left behind it, which is called a hole, and in the appearance of a negative charge carrier in the conduction band, this being the electron. With the return of the electron to the valence band, which can occur as a result of spontaneous or induced transition, a reverse process takes place, namely, a recombination of the electron-hole pair, accompanied by the emission of energy in the form of a quantum of light radiation.

In a pure (so-called intrinsic) semiconductor the number of free electrons and holes is the same, being dependent only on the temperature: the higher the temperature, the greater the number of current carriers. However, the concentration of such current carriers in an intrinsic semiconductor is relatively small. Non-equilibrium conditions are absent, since the number of particles in the upper energy level cannot be greater than the number of particles in the lower energy level. Therefore non-equilibrium conditions are created by using special methods.

The concentration of current carriers in an intrinsic semiconductor can be increased by doping it with special impurities that have an electron or hole conductivity. These impurities introduced into different parts of a semiconductor form conductivity regions of two types, one of which is called an *electron conductivity region* (n-region) and the other a *hole conductivity region* (p-region). Such a semiconductor is called extrinsic. By applying an electric field to an extrinsic semiconductor, it is possible to make the electrons and holes in it move towards each other. In a small part of the extrinsic semiconductor crystal, which corresponds to the transition from the n-region to the p-region (called a *p-n transition* or *junction*), the concentration of electricity carriers becomes very high, non-equilibrium conditions are created, holes and electrons undergo recombinations with the emission of quanta of electromagnetic energy (Fig. 47).

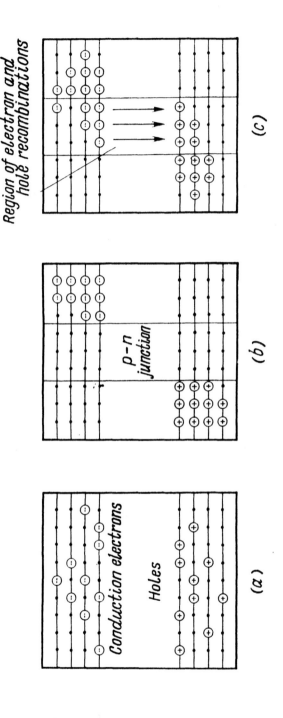

Fig. 47. Distribution of electrons and holes in a semiconductor

(a) at a temperature above absolute zero; (b) current is not supplied to the p-n junction; (c) current flows through the p-n junction, electrons and holes undergo recombination, radiation takes place at high current values

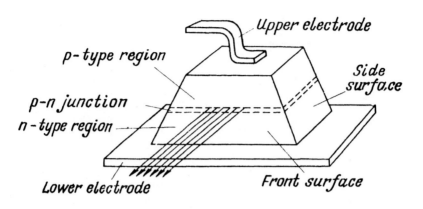

Fig. 48. Design of a semiconductor laser emitting element

In the first models of the semiconductor laser the active medium was a single crystal of gallium arsenide GaAs cut into a platelet having a thickness of only 0.5 mm. (Fig. 48). This platelet is not homogeneous: it consists of two parts exhibiting an electron conductivity and a hole conductivity respectively. The thickness of the *p-n* junction layer is as small as some thousandths of a centimetre, but it is in this layer that the emission is stimulated. Electric current is applied to the crystal platelet through a strip electrode fixed to its upper surface.

With the current supplied from a suitable source, the stimulated emission is propagated in the plane of the *p-n* junction. Photons emitted at the moment of recombination of an electron with a hole will stimulate recombination of other carriers of electric charges. The result will be stimulated emission of radiation. The frequency of this radiation is determined by the width of the bandgap. If the two opposite faces of the crystal, lying perpendicular to the boundary of the *p-n* junction are made strictly parallel, ground and polished to a high precision, the crys-

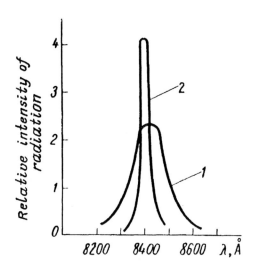

Fig. 49. Spectral characteristics of semiconductor laser radiation

1—with low current density; *2*—with high current density.

tal will be a resonator which at a sufficient current density will be capable of generation.

The first models of gallium arsenide lasers generated in a pulse mode only at current densities of the order of 8000 A/cm^2 and low temperatures corresponding to that of liquid nitrogen. The duration of a pulse was from 5 to 20 μs. Usually, however, in semiconductor lasers the threshold current density ranges from 700 to 20 000 A/cm^2 and is determined by the pulse duration and semiconductor temperature.

When exciting currents are small, only a small part of carriers undergo recombination, and the radiation process is spontaneous. The laser radiation is therefore random, incoherent and broad-band, with a low intensity (Fig. 49). With an increase in the current density the emission becomes more and more coherent, the spectral linewidth sharply decreases, the radiation intensity markedly increases, and the emergent beam narrows.

The first gallium arsenide semiconductor laser generated at 8400 Å, its linewidth being several angstroms and beam divergence in the plane of the *p-n* junction, 4°.

One of the first lasers employing gallium arsenide doped with zinc and tellurium and working pulsed at room temperature gave radiation with a power of several tens of watts. When cooled with liquid neon (27°K), the laser operated CW, its output power being up to 3.1 W and efficiency, 50 per cent.

By now, besides gallium arsenide lasers, several lasers using other semiconductor crystals have been developed. Such are a semiconductor laser with gallium arsenide-phosphide (GaAs-P) as the active medium, having the radiation wavelength of 0.65 to 0.85 μ and working pulsed at the temperature of liquid oxygen; an indium phosphide (InP) laser with the radiation wavelength of 0.91 μ, working pulsed at 90°K and CW at 20°K; an indium arsenide (InAs) laser with the wavelength of 3.1 μ, working pulsed at 77°K and CW at 40°K.

There are many other semiconductors which may be used as active media for obtaining coherent radiation. It is possible to create lasers on the basis of mixed semiconductor crystals, which would ensure generation at all wavelengths from the red, visible region of the spectrum to 5-10 μ.

In the last few years, besides the method of direct electric excitation, termed injection, other methods of obtaining inverted population have also come into use. Thus if a semiconductor is subjected to bombardment with a bunch of electrons possessing a sufficient energy (of the order of 20 keV and higher), such as is formed in electron-beam devices, coherent radiation can also be generated.

Electrons, possessing a large amount of kinetic

energy, penetrate into the semiconductor and in the path of their travel ionise its atoms. The electrons which have formed as a result of this ionisation, when colliding with the atoms of the semiconductor crystal lattice, excite other electrons and bring them from the valence band to the conduction band. Electron bombardment causes avalanche ionisations in the semiconductor. Each ionisation act is accompanied by the appearance of a pair of carriers—an electron and a hole. Each exciting electron of the bunch creates up to several tens of thousands of electron-hole pairs in the semiconductor. Giving off the excess energy to the crystal lattice, the charge carriers accumulate in energy levels near the edges of the valence and conduction bands. From these states the electrons and holes can recombine with the emission of light quanta. Induced emission emerges from the semiconductor through its reflecting side surfaces in a direction perpendicular to the incidence of the electron beam (Fig. 50).

This excitation method is advantageous in that a considerable depth of penetration of exciting electrons allows the obtaining of inverted population in greater volumes of the active medium and, hence, makes possible the creation of more powerful semiconductor lasers. A disadvantage of this method resides in that with electron-beam pumping radiation can be obtained only at low temperatures. The efficiency of such lasers does not exceed 12 per cent.

Recent laboratory investigations have shown that semiconductor optical generators can be pumped in the same manner as solid-state ones, i.e. by using light. Photons possessing an energy greater than the width of the bandgap, on being absorbed in the semiconductor, are capable of transferring electrons from the valence band to the conduction band, thus ensu-

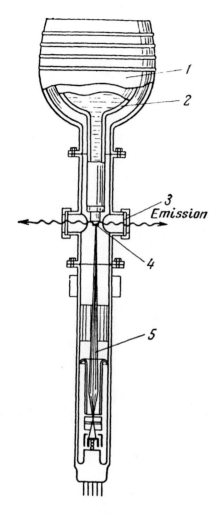

Fig. 50. Design of a semiconductor laser with electron-beam pumping

1—dewar; 2—liquid nitrogen or helium; 3—exit window; 4—semiconductor crystal; 5—electron beam

ring inverted population and, consequently, conditions required for the generation. Optical pumping can be performed by means of an injection laser or a ruby laser. However, the efficiency of optically pumped semiconductor lasers so far remains insignificant (not over 1 per cent).

Finally, generation of charge carriers and inverted population in semiconductors can also be achieved with the aid of a strong electric field. Moving in an accelerating electric field, free carriers acquire considerable energy and, colliding with the semiconductor atoms, can ionise them. This results in a higher concentration of electrons in the conduction band and holes in the valence band, so that an inverted population state can be set up. Direct electric excitation was attained only in one material—gallium arsenide (GaAs). Recombination in the semiconductor

Table 4

Semiconductor material	Wavelength, μ	Method of excitation
Gallium arsenide, GaAs	0.85	Injection
Indium phosphide, InP	0.9	Same
Gallium antimonide, GaSb	1.6	Same
Indium arsenide, InAs	3.2	Same
Lead sulphide, PbS	4.3	Same
Indium antimonide, InSb	5.3	Same
Lead telluride, PbTe	6.5	Same
Lead selenide, PbSe	8.5	Same
Zinc sulphide, ZnS	0.33	Electron-beam
Cadmium sulphide, CdS	0.5	Same
Cadmium selenide, CdSe	0.69	Same
Cadmium telluride, CdTe	0.8	Same
Gallium arsenide, GaAs	0.85	Same
Gallium antimonide, GaSb	1.6	Same
Cadmium sulphide, CdS	0.5	Optical
Gallium arsenide, GaAs	0.85	Same
Indium arsenide, InAs	3.2	Same
Lead telluride, PbTe	6.5	Same
Gallium arsenide, GaAs	0.85	Ionisation

takes place only after the removal of the external electric field. Therefore such a laser is capable of pulsed working only.

Semiconductor lasers are finding an ever increasing application in laboratory research, in communication, in high-speed computers, in laser ranging, in medicine, and in biology, since these devices feature wide frequency spectrum, are compact, capable of inertialess operation and can be excited by various methods.

Fig. 51. Semiconductor optical quantum generator "Mayak"

Shown in Fig. 51 is a semiconductor optical quantum generator "Mayak" which gives coherent infra-red radiation at the wavelength of 0.9 μ. It can be employed for investigating emission under various atmospheric conditions, for information transmission, for measuring distances, for remote control of mechanisms, for scientific research in biology and other branches of science and engineering. The output power per pulse is 6 to 8 W, with the pulse duration being 0.15 μs. The weight of the laser is 1.5 kg.

The main characteristics of semiconductor lasers are given in Table 4.

CHAPTER 4

Application of Lasers

LASERS IN COMMUNICATIONS

The radio-frequency range has become so "crowded" that scientists were to tackle the problem of mastering a new range. But wavelengths of some decimal fractions of a millimetre turned out to be the limit attainable with conventional radioengineering methods in the present state of technology.

The mastering of the optical range opens new prospects for communications. The entire range used for radio communications is known to occupy the frequency band of approximately from 10^4 to $3 \cdot 10^{11}$ Hz, while the optical range extends from $3 \cdot 10^{12}$ to $15 \cdot 10^{15}$ Hz. Simple calculations show that the optical range is approximately 50 000 times wider than the radio range. Using a laser beam as a communication link, it would be possible to transmit hundreds of thousands of television programmes or ensure simultaneous telephone conversations for the entire population of our planet. Calculations show that with relatively small powers it is feasible to communicate in the outer space over such distances which only yesterday seemed fantastic. But communications with the help of lasers can not only broaden the potentials of radioengineering means. Such techniques can aid and, where necessary, replace conventional telephone communications. Successful experiments in this direction indicate that practical implementation of lasers in telephone communications is well in sight.

For the development of real laser communication systems under the conditions of the terrestrial atmosphere, it is necessary to take into account the characteristic properties of the propagation of the electromagnetic oscillations of the visible and infra-red regions in it.

Gases, minutest solid and liquid particles suspended in the atmosphere can substantially affect the propagation of light. The optical properties of the atmosphere are influenced mainly by carbon dioxide gas, ozone and aerosols (dust, smoke, droplets of water, small crystals of ice, etc.). The layers in the atmosphere are constantly intermixing. In the lower layers of the atmosphere the main quantity of water is concentrated, coming to 4 per cent of the atmosphere volume, in the form of minutest drops, mist, and vapours. The propagation of light is influenced by atmospheric precipitations. Rain and snow strongly affect the transparency of the atmosphere.

Fog is very hazardous for reliable optical communications. The period of heavy fogs, rains and snowfalls during a year depends on the geographical region, season and time of day. In Moscow, for example, heavy fogs average 70 to 80 hours a year.

All these factors impose specific requirements on the organisation of communication links using laser systems. It is obvious that for the communications to be stable and high-quality under any meteorological conditions, the influence of these adverse factors must be minimised.

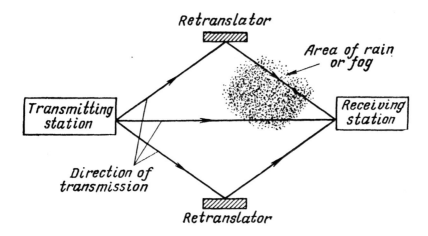

Fig. 52. Scheme of effecting simultaneous transmission in several directions

For raising the reliability of the communication system it is necessary to properly select its operation frequencies since the passage of light vibrations having different frequencies through rain, fog or snow regions is not the same.

Another way of overcoming these difficulties and increasing the reliability of communication can be the development of a system with the transmission doubled simultaneously in several directions (Fig. 52).

Atmospheric absorption imposes a limitation on the use of lasers for terrestrial communications. Nevertheless, even now with the help of gas lasers information can be transmitted over distances of dozens of kilometres in fine weather at night and of about some ten kilometres in case of thin fog and rain. If the emitters and receivers are placed at an altitude of 60 to 100 m, the adverse influence of the fog will be drastically diminished and the communication coverage can be extended.

A further increase in the power of the laser and sensitivity of the receiver will allow an extention of the transmission coverage under any meteorological conditions.

LIGHT BEAM MODULATION METHODS

The most important engineering problem in the creation of laser communication lines and in the transmission of signals over such lines is the modulation of the light beam. There are many physical phenomena on which different methods of modulation are based.

The choice of the light beam modulation method is dictated by the requirements to the intensity of the luminous flux, maximum modulation factor and work of the laser (whether it operates by emitting individual pulses or trains of pulses).

None of the existing modulation methods can be considered universal: in each particular case its choice lies with specific requirements to be met.

The simplest modulator can be a rotary disk with radial slots. Such a disk arranged in the path of a laser beam will periodically interrupt the luminous flux. An advantage of such a modulator is that the laser radiation frequency has no influence on the modulator characteristics, so that it can ensure a 100 per cent modulation. The modulation frequency is determined by the number of slots in the disk, by the speed with which it rotates, and reaches several megahertz.

This method of modulation cannot be employed for transmitting information, but if the slots in the disk are of different transparency and calibrated to a high degree of precision, then such a rotary disk can function as an optical attenuator.

Modulation can also be effected by means of a rotary mirror. In this case use is made of an octahedral mirror and the so-called image dissection method is employed. Two gratings are employed for the purpose. The image of one grating is superimposed on the image of the second grating. The speed with which the image of one grating shifts in relation to the surface of the second grating is determined by the rotation speed of the mirror, the distance between the gratings and the distance by which these gratings are spaced from the mirror. With this method the modulation frequency is of the order of 100 MHz (Fig. 53).

Another method of modulation is based on varying the transparency of a prism by bringing another auxiliary prism close to the reflecting surface of the first prism (Fig. 54). Piezoelectric and magnetostrictive crystals are employed as an auxiliary prism.

Modulation can also be performed by varying the quality factor of the laser cavity. In this case the modulating element is arranged between the end face of the ruby rod and the surface of the reflector, the active zone thus being as if increased. Such modulating element is a crystal with electrooptic properties. By applying a certain voltage to it and varying this voltage, it is possible to vary the optical path length of the resonant cavity, its quality factor and the intensity of the laser output signal, i.e. to effect its modulation.

But most effective and, probably, most promising are polarisation methods of modulation.

As is known, a light beam can be polarised, that is, there can be obtained a light wave with a definite orientation of its electric and magnetic vectors. Ordinary light consists of light waves whose oscillations may occur in any direction whatsoever. Polari-

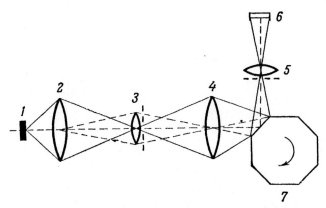

Fig. 53. Scheme of an image dissector with a rotating mirror
1—light source; 2,4—lens; 3,5—grating; 6—detector; 7—octagonal mirror

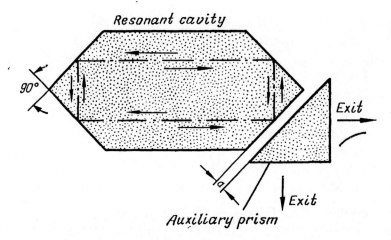

Fg. 54. Scheme of modulation based on varying the transparency of a prism. Radiation intensity is varied by varying gap "a"

sed light consists of light waves whose oscillations occur only in one quite definite direction.

Light can be partially polarised as well. Such light can be regarded as a mixture of ordinary and polarised light.

Polarisation phenomenon is observed, for example, when light passes through a tourmaline plate. The ray of light which emerges from such a plate is plane polarised. Now, if a second tourmaline plate is interposed in the path of the ray emerging from the first tourmaline plate, then, depending on the orientation of the second plate in relation to the first one, the intensity of the light ray can be modified till its complete extinction. The polarisation phenomenon is also observed in case of reflection or refraction of light at the boundary between two isotropic dielectrics.

Double refraction is displayed by some crystals, for instance by Iceland spar which is a variety of calcium carbonate $CaCO_3$. If one takes a crystal of Iceland spar and looks at a printed page through it, all the letters will be doubled. This phenomenon is caused by two mutually perpendicular rays emerging from the Iceland spar crystal, one of these being called the *ordinary ray* and the other, the *extraordinary ray*.

The most commonly used polariser is a Nicol prism, often referred to simply as a *nicol*. The Nicol prism is a crystal of Iceland spar sliced from the one blunt corner to the other in a plane parallel to the long diagonal of the end faces; the cut faces are re-united with a film of Canada balsam. One of the rays originating in the Nicol prism as a result of double refraction is eliminated in a rather ingenious manner. Since the ordinary ray suffers a stronger refraction, it falls on the boundary with Canada balsam at an angle greater than the angle of incidence of the extraordinary ray. The refractive index of Canada balsam being lower than that of Iceland spar, a total internal reflection takes place, and the ray falls on the side face which is painted black and therefore completely absorbs the incident ray. As a result, only one plane-polarised (extraordinary) ray emerges from the prism. The plane of polarisation of this ray is called the *principal plane of the nicol*. Two nicols arranged so that their principal planes are at right angles to each other do not transmit light. In case their principal planes are parallel, light will pass through freely. With any other mutual arrangement of the principal planes, light will pass through the nicols only partially. With the principal planes arranged parallel, maximum amount of light will be transmitted through them.

The above-considered effect, as well as the well-known Kerr, Pockels and Faraday effects, are employed for the modulation of the outgoing laser beam. Let us now briefly discuss how amplitude modulation of light is performed on the basis of these effects.

The Kerr effect consists in that certain transparent liquids, when subjected to the action of an electric field which creates a structure similar to crystalline, as a result of a definite orientation of their molecules, become doubly refracting.

If we take a vessel with transparent plane-parallel walls, fill it with nitrobenzene (which is an electrically non-conducting liquid), immerse two plate-shaped electrodes into it, apply a high voltage across the electrodes, and then pass polarised light between them, the result will be the Kerr effect. The arrangement by means of which it is produced is a Kerr cell. The phenomenon observed at the cell output will be double refraction.

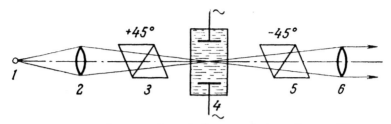

Fig. 55. Scheme of modulator with a Kerr cell

1—light source; *2*—objective; *3*—polariser; *4*—Kerr cell; *5*—analyser; *6*—lens

The shift between the ordinary and extraordinary rays passing through the liquid is proportional to the distance travelled by the light in this liquid, to the square of the electric field strength and to a certain constant characteristic for a given medium. The Kerr effect is displayed by many liquids, but it is most pronounced in nitrobenzene, for which reason in modulators preference is now given to this compound. Figure 55 is a schematic diagram of a modulator with a Kerr cell.

The beam from a light source located at point *1* passes through condensor *2* and further through polariser *3* which is a Nicol prism. Here the laser beam becomes linearly polarised. Then the light beam focused in the interelectrode space passes through Kerr cell *4* to the electrodes of which a voltage is applied, and becomes elliptically polarised. If the applied voltage is variable, the direction of the beam polarisation will also vary in accordance with the value of the voltage applied. With no voltage across the plates of the Kerr cell the system will be non-transparent.

Then the beam is directed to a second polariser *5* which is referred to as an analyser, the optic axis of this second polariser being at right angles to that of the first polariser.

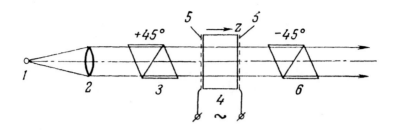

Fig. 56. Scheme of modulator with a Pockels-effect cell
1—light source; *2*—objective; *3*—polariser; *4*—crystal of active material; *5*—electrodes; *6*—analyser

The advantage of this method consists in that it is practically inertialess. The Kerr effect has rather long been known to be used in engineering. For instance, one of sound recording systems employed in cinematography is based on it.

This is the least complicated and so far preferred method of modulation. It can be practised for modulating beams in the visible and near infra-red regions of the spectrum.

Another kind of a modulator employs the Pockels effect. The Pockels effect is displayed only by piezoelectric crystals. These crystals do not split a light beam into the ordinary and extraordinary rays along one of the axes (Z-axis), but they do along the others. But if an electric field is applied to the crystal along its Z-axis, then the optic axis will be split into two and the crystal will become biaxial. A light beam will be split into the ordinary and extraordinary rays. This phenomenon is called the Pockels effect. At present crystals of ammonium dihydrophosphate and potassium dihydrophosphate are widely used for obtaining this effect.

A schematic diagram of a modulator operating on the Pockels effect principle is shown in Fig. 56.

Light from source *1* is passed through objective

2 and, after having traversed polariser *3* falls as a parallel beam upon crystal *4* parallel to its Z-axis. Modulating voltage is applied to the crystal by means of grid electrodes *5*. When analyser *6* and polariser *3* are crossed and the system is de-energized, it does not transmit the light. With voltage fed to electrodes *5*, the crystal becomes doubly refracting in the direction of Z-axis. The plane of polarisation is rotated and the system transmits light in accordance with the voltage impressed. Modulation is thus effected. Unlike the case of the Kerr cell, the light beam here must be parallel to the Z-axis. The modulator can operate within the entire visible range and in the near infra-red region of the spectrum.

There is still another kind of modulator, whose operation is based on the Faraday effect. The Faraday effect consists in that in some optically active media the plane of polarised light is rotated under the action of a magnetic field. In this case the direction of light must coincide with the direction of magnetic lines of force. The rotation of the plane of polarisation depends on the optical path length in the active medium, on the magnetic field strength and on a certain constant for a given medium. The Faraday effect is displayed by some solids (quartz, glass) and liquids (carbon disulphide, petrol, water).

A schematic diagram of a modulator in which the Faraday effect is employed is shown in Fig. 57. Light from source *1* through objective *2* and polariser *3* passes as a parallel beam through active material *4* in the magnetic field of solenoid *5*. Analyser *6* is arranged in such a manner that its plane of polarisation should make an angle of 45° with the plane of polariser *3*. In case the voltage applied to the solenoid is zero, the transmission coefficient of the modulator is 0.5. With the applied voltage varying

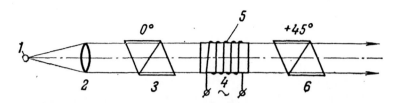

Fig. 57. Scheme of modulator based on Faraday effect

1—light source; *2*—objective; *3*—polariser; *4*—active material; *5*—solenoid; *6*—analyser

from positive to negative, the direction of the magnetic field of the solenoid will also vary, which, in turn, will determine the rotation of the light beam polarisation plane. The luminous flux at the output of the system will be changed.

The polarisation method of light modulation can be realised resorting to the Pockels effect. While with amplitude modulation the result was a modulated light signal whose amplitude (i.e. the magnitude of the luminous flux) changed in accordance with the modulating signal, with the polarisation method only the polarisation plane will change in accordance with the modulating signal, the luminous flux remaining the same. This change of the polarisation plane will be detected at the receiving end by a demodulator made as an analyser. A modulated signal will be obtained at the output.

This method is most promising for terrestrial communication lines, i.e. under atmospheric conditions. One of its disadvantages lies with polarisation distortion because of atmospheric turbulence.

Besides the amplitude modulation of light, frequency and phase modulations of light can be performed as well. In other words, it is possible to effect any type of modulation, which is based on varying one of the characteristics of the light wave (or several such characteristics simultaneously).

Frequency modulation of light is based on the Zeeman effect. When the active material of a laser is subjected to the action of a magnetic field, one of the spectral lines of the laser radiation may be split into two. This brings about the origination of the so-called super-fine structure of the spectral lines, which is caused by a change in the projection of the electron orbital moment onto the vector of the external magnetic field. In other words, new levels appear in the energy spectrum of the atoms. With a variation of the external magnetic field, the frequency of radiation which takes place as a result of transitions from these new levels also varies in accordance with their energy.

Frequency modulation of light can be effected by using a tunable Fabry-Perot interferometer. As is known, the interferometer transparency changes for each wavelength as light passes from one reflecting surface to the other, this transparency depending on the distance between the reflecting surfaces. Therefore, if the distance between the reflecting surfaces is varied, the frequency of the laser operation will vary accordingly. Hence, frequency modulation can be achieved by fixing one of the mirrors on a piezoelectric plate whose thickness can be varied depending on the voltage applied to it.

The same principle is employed for amplitude modulation, though frequency modulation should be preferred as ensuring better signal-to-noise ratio under otherwise equal conditions.

Since laser output beams are coherent, phase modulation is not difficult to effect either. To do this, the light beam must be passed through an electrooptical medium subjected to the action of a modulating electric field. At the output of such a modulator the light wave of the extraordinary ray will be phase-modulated because the propagation velocity of the extraordinary ray in the electrooptical medium differs from the initial phase velocity of this ray. To detect the phase-modulated oscillations, they must be compared with a reference frequency signal. For obtaining this signal, part of the output energy of the quantum generator can be directed past the modulator and then mixed on a photodetector with the light signal which has undergone phase modulation.

Experimental investigations of this method of modulation have shown it to be most promising. The phase modulator is almost insensitive to internal stresses and optical adjustment inaccuracies.

It should be pointed out that though not much time has passed since quantum generators were created, many other methods of modulation besides those mentioned above, such as pulse, pulse-code, single-sideband modulation methods, have been suggested and realised.

In receiving communication systems with the use of lasers electric photodetectors are employed. The principle on which they operate is the conversion of radiation energy into electric energy due to internal or external photoeffects. Photodetectors whose operation is based on the external photoeffect are photocathodes, photoelectron multipliers, and those whose operation is based on the internal photoeffect are photoresistors and photodiodes.

For optical communication systems with a wide-band modulation preference is given to travelling-wave tube photodetectors and photoparametric diodes. One of the possible particular designs of optical telephone communication system is shown schematically in Fig. 58.

The sharing equipment of two central offices is interconnected by optical range beam waveguides.

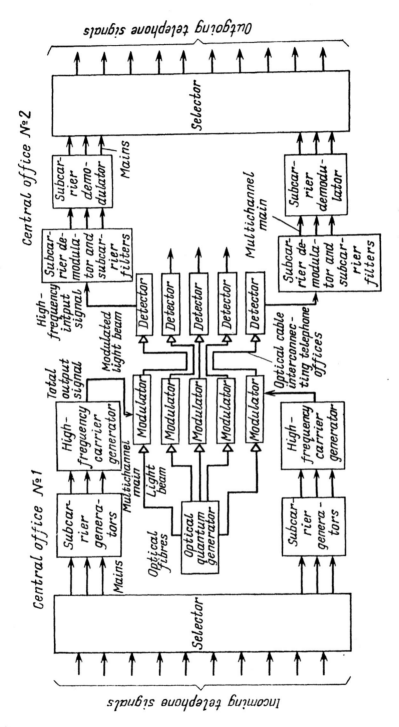

Fig. 58. Version of a laser telephone communication system

A subscriber's call signal comes to a selector which performs hunting and offers a free main. Then the message signal passes through a number of high-frequency modulators and a group signal is shaped. Such group signal may contain the information pertaining to hundreds and thousands of conversations carried out simultaneously. The group signal further comes to a next modulator where it modulates the laser light beam arriving at the modulator along an optical waveguide.

On leaving the modulator, the modulated light beam enters an optical waveguide and is transmitted along this waveguide to the second central office. The process which takes place at the second central office is opposite to that at the first central office, and it results in restoring the initial speech signal at the output of the sharing equipment, this restored signal being then sent to the second subscriber.

The system just described is only one of the possible versions of realising optical telephone communications with the use of lasers.

BEAM WAVEGUIDES

Limitations imposed on the applicability of laser beams for communication purposes in the lower atmospheric layers call for the creation of some kind of protected long-path media for the propagation of light.

For instance, an ordinary tube or pipe can be used for transmitting light through it, by making the laser beam propagate along this pipe. Since the beam divergence is small, the length of the pipe could be 5 km or so. But actually it is not at all easy to make such a pipe straight, even if it is laid on the ground or rests on special supports. The light beam will inevitably strike against the inner surface of the pipe, which will bring about substantial transmission losses and considerable phase distortions.

It is possible to manufacture pipes (and attempts have been made in this direction) with a very precise boring and a mirror finish of the internal surface. Light, while propagating along it, undergoes multiple reflections.

Taking into account the possibility of creating optical communication systems with a high degree of sharing (millions of simultaneous telephone conversations, thousands of television programs transmitted via a single beam), attempts are made at developing an effective method for transmitting the radiant energy of a laser along a waveguide. In case of success, any complexity of the design of such a waveguide would very soon be justified.

At present three types of beam waveguides are under consideration: diaphragmatic waveguides, waveguides employing dielectric lenses, and waveguides in which gas lenses are used.

A diaphragmatic waveguide is a tube with diaphragms mounted inside it on stable supports equally spaced from one another (Fig. 59a). The light beam diameter in such a waveguide is somewhat restricted by the diaphragm aperture. The resulting distortions are compensated for as the beam passes from one diaphragm to another.

When the aperture is large as compared to the wavelength, the losses will be small. Thus, assuming the wavelength λ to be 1 μ, the space between the diaphragms $D = 10$ m, and the radius of the diaphragms $R = 1.7$ cm, the losses will be 1 db/km only.

An important step in tuning the diaphragmatic waveguide is the adjustment of its apertures. High sensitivity of the diaphragmatic waveguide to the adjustment of the apertures has been confirmed by the experiments. The waveguide diameter must be such that the light beam could negotiate the bends unobstructed.

A waveguide with dielectric lenses (Fig. 59b) has lens-shaped phase correcting plates. Each subsequent lens restores the phase distribution along the beam cross section, which existed immediately after the previous lens. Diffraction at the lens aperture exerts but small influence on the passage of the beam, though is a source of losses.

As in the case of a diaphragmatic waveguide, losses may be very low, if the aperture is large as compared to the wavelength. Thus, a lens waveguide having the same distance between the lenses as that between the diaphragms in a diaphragmatic waveguide ($D=10$ m), with the wavelength $\lambda=1$ μ, can be made with the aperture radius of only 3 mm. Since such a size is very small, it is reasonable to set another distance between the lenses. If this distance $D=100$ m, then the radius should be increased to 10 mm. In such a case the diffraction losses should be of the order of 0.01 db/km.

The accuracy of the adjustment of the lenses here may be not so high as with diaphragms. It should be pointed out in this connection that small curvatures of the lenses or a variation in the distance between them have little effect on the losses. Naturally, sharp turns should be made with the help of prisms or mirrors.

Such a lens waveguide will possibly be less expensive to manufacture than a diaphragmated one and work better.

A special mock-up was constructed with a view to answering a number of problems associated with practical realisation of a laser beam transmission

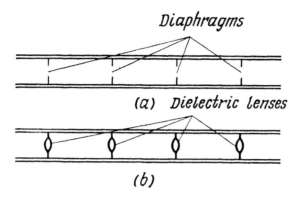

Fig. 59. Waveguides

along a waveguide employing dielectric lenses. The waveguide was an aluminium pipe having 102 mm in diameter. To preclude temperature fluctuation and turbulence effects in the path of the light beam, this pipe was placed into another aluminium pipe with a diameter of 152 mm. The external pipe was mounted on conventional pole supports. The length of all the pipe sections taken together was 970 m. Lenses were spaced 97 m apart and their focal length was about 48 m.

In this experiment the source of light waves for the waveguide was first a mercury lamp, and then a conventional tungsten incandescent lamp. No continuous-wave lasers existed at the time of this experiment. A collimator lens with the focal length of 1 m and an optical filter for passing only part of the spectrum was arranged in the path of the light. The light source was modulated by a mechanical interrupter with a frequency of 1000 Hz.

The results of the experiment showed the beam passage to be considerably influenced by the air motion and thermal stratification in spite of the double screening, and this presents difficulties in the adjustment of the lenses. The output signal was stable only at night when temperature fluctuations were small. The level of the signal at the line output proved to be lower than expected.

The experiment is planned to be repeated with greater attention being paid to the quality of the lenses. In this new experiment it will be possible to use a CW gas laser instead of an incoherent source. With a CW gas laser measurements can be conducted both in the optical and infra-red regions. A new method with the use of gas lenses, which appears to have considerable promise, has been developed in the USA. The operation principle of gas lenses is based on temperature variation of the refractive index of gas. In contradistinction to the above-considered methods of transmitting the energy of a laser beam along waveguides, gas lenses practically neither reflect nor absorb transmitted light.

A narrow laser beam can be made to be directed along the axis of the pipe in case a long gas lens or a series of such lenses are employed. These lenses do not possess any considerable power; they only compensate for the natural divergence of the laser beam within rectilinear sections of the pipe. Where the pipe is bent, the light beam, getting into the area with a lower refractive index, is deflected towards the area with a higher refractive index. The sharper the bend, the stronger the beam focusing should be. Thus a pipe whose bends follow the variations of the Earth's surface can be used as a laser long-range communication line.

The principle of operation of gas lenses is based on the well-known phenomenon consisting in that light beams are refracted towards a medium having a higher refractive index. On account of this property the light beam can be focused axially along a region which has a higher refractive index than the surrounding medium.

One of the possible designs of a gas lens is shown in Fig. 60a. Arranged along the axis of a gas-filled pipe is a spiral through which electric current is passed for heating the gas. In the direct vicinity of the spiral turns the gas is heated to a greater extent than inside the spiral. The refractive index of the gas is proportional to its density. This allows the focusing of the light beam.

A mock-up of a gas lens was constructed. The experimental gas lens was a small section of a pipe with a spiral, 75 cm in length. The pipe was filled with different gas-and-air mixtures. The possibility of using carbon dioxide, freons and certain hydrocarbons for this purpose was investigated. The gas lens was adjusted by varying the temperature of the spiral. No aberration was observed with the focal length over 5 m.

Besides the lens of the type discussed, it is possible to use counterflows of two gases which have different optical densities and are admitted into a mixing chamber (Fig. 60b). The two gases entering the chamber are mixed in it and then the mixture is withdrawn from the chamber. The rate of flow of the gases and the geometry of the mixing chamber are selected such that a symmetrical boundary layer should be formed in the region where the two gas flows meet. A light beam is focused as it passes through such a mixer.

The model of the lens just described was tested with flows of argon and carbon dioxide having the same temperature. As with thermal gas lenses, no aberrations were observed for the same convergence factors. All the above considered types of waveguides offer an essentially new approach to the problem of

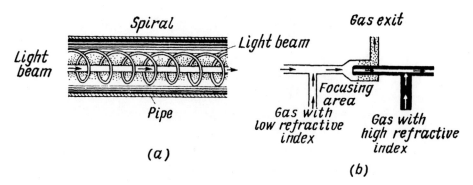

Fig. 60. Design of gas lenses
(a) convection lens; (b) mixing lens

transmitting light signals over large distances, though further research is needed before they can be practically employed.

It is not improbable that soon we shall see beam waveguides being laid instead of telephone communication cables and radio relay lines, which will serve both local telephone exchange and long-range communication purposes, replace former small-capacity facilities and ensure the transmission of any kind and amount of information between any localities on our planet.

LASERS IN COMPUTERS

Improvements in modern electronic computers are directed towards increasing their high speed and reliability. To this end, literally all the achievements made in present-day physics are used: magnetic properties of films, parametric oscillations, tunnel effect, etc. Increase in the high speed of the computer operation goes hand-in-hand with miniaturisation of computer components. But the challenge faced here was the mutual interference of conductors which behave as radiating aerials and induce noises in the near-by elements.

Therefore the appearance of lasers at once suggested the use of their principles in computer designs. This idea, when realised, would solve two problems simultaneously: increasing of the high speed and elimination of mutual interferences.

Any modern computer consists essentially of logical circuits, a memory (storage) system, and means for transmitting the information being processed. It appears feasible to create electronic computers built around lasers with a circuitry based on a radically new principle.

Quite a number of investigations have been conducted in this direction, though it would be premature to draw any final conclusions. One thing is obvious, however: with optical transmission lines the wavelength of signals is many times less than the dimensions of any circuit elements and therefore mutual parasitic influences can be obviated.

The possibility of transmitting signals between individual components of the computer system without recourse to any contacts opens new vast prospects for the design of computer elements. The use of optical signals offers an absolutely new approach to the construction of information transmission circuits.

Optical computers on lasers will have very high operational velocities, much superior to those of which the now existing electronic devices are capable. The information processing capacity of future systems will be greatly increased.

For transmitting light pulses between the internal elements of an optical computer, materials can be employed in which the propagation of light suffers only a small attenuation. Glass fibres noted for very insignificant losses are of greatest interest as a material for such conductors. By coating such optical fibre with a thin layer of glass featuring a lower refractive index than that of the fibre, it is possible to eliminate completely the mutual interference between two neighbouring transmission lines.

Optical fibres used as light guides can be very thin. Thus a bunch that is 10 μ in cross section consists of up to 100 light guides.

A computer on laser neuristors is planned to be created in the United States. All signals, whether information or control ones, will be optical. The main elements of the computer will be made from glass fibre with a definite concentration of active (radiating) and passive (absorbing) ions. The computer will be powered from a continuous light medium to ensure constant pumping power for maintaining inverted population of the radiating ions.

The principal advantages of such a system will be: no need for connecting wires to power individual circuits of the computer, possibility of transmitting signals without any auxiliary connectors, and great high-speed potentialities.

Other suggestions have also been made for the realisation of various optical computer designs. In the creation of logical circuits or memory systems on lasers use can be made of their optical interaction. The light emitted by one semiconductor laser can be extinguished by the coherent light of the other laser. The same phenomenon is observed with neodymium glass lasers. The essence of this phenomenon is as follows.

Suppose we have a system of two lasers and the directions in which the beams are formed in their respective active media are mutually perpendicular. The active material of each laser is square in section.

If we pass coherent radiation of one of the lasers through the preliminarily excited active medium of the other laser, radiation will set up in the latter, the direction of this radiation being the same as in the first laser. The energy of the excited particles of the second laser will be given away to the light beam emitted by the first one. Naturally, the second laser will then be unable to generate on its own. If the two lasers are of the same power, the beam of the first laser, when emitted, totally extinguishes the beam of the second laser. A system of two lasers which mutually extinguish each other and have extinction coefficients somewhat greater than unity can be shown to possess two stable working states *1* and *2*, i.e. to be a bistable system. The state when the laser radiates will be referred to as state *1*, and the state when it does not radiate, state *2*.

The extinction coefficient should be understood as the ratio of the power taken away from the laser being extinguished to the power radiated by the extinguishing laser.

In the system shown in Fig. 61 two similar semiconductor lasers A and B are employed, each being square in cross section. One side of the square is an ideal reflecting mirror, and the side opposite to it, a partially reflecting mirror. The two other sides of the square are wholly transparent. Coherent light emerging from laser A passes through laser B crossing the non-reflecting planes of the latter and therefore the two lasers are not embraced by a feedback.

The simplest symmetrical arrangement of two lasers in which they can extinguish each other is shown in Fig. 62. In this scheme absorbing sections or focusing can be employed. But such a scheme is disadvantageous in that the optical distance between the elements cannot be smaller than the length of

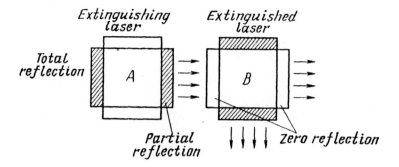

Fig. 61. Extinguishing of semiconductor lasers

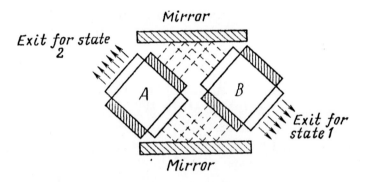

Fig. 62. Bistable two-laser system

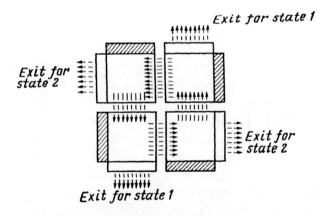

Fig. 63. Bistable four-laser system

the laser side. Hence, for limiting the diffraction divergence of the beam, the dimensions of the both lasers must be minimised. This difficulty is overcome in an arrangement using four appropriately positioned lasers. The scheme of such an arrangement is shown in Fig. 63. No additional mirrors are needed in it.

Under extinction conditions, the power of the emerging beam is approximately twice that of the induced radiation of a single laser. Therefore, an apparatus employing two lasers (Fig. 62) can switch over two similar devices when it splits the beam; in other words, the multiplication factor for these apparatus is 2. In the arrangement shown in Fig. 63 four similar devices can be switched over.

Apparatus on lasers built in accordance with these schemes may find application in logical and memory systems. In such systems light pulses produced by semiconductor lasers are the sole carriers of information. The function of electric current here is confined to the excitation of optical quantum generators.

APPLICATION OF LASERS IN METROLOGY

Optical quantum generators can be used in metrology. The conventional techniques of measuring lengths with an accuracy of fractions of a micron, based on interference phenomena are extremely complicated and not fit for measuring lengths exceeding one metre.

As is known, the standard of length adopted at present is the wavelength of the orange line emitted by the isotope of krypton with the mass 86. In one metre this length is contained 1 650 763.73 times. By using an interferometer with a krypton lamp, the length of any specimen can be measured with a high degree of precision. But if the specimen has a length

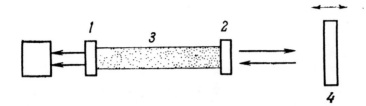

Fig. 64. Measuring lengths by means of a laser

1,2—mirrors of a laser resonator; *3*—gas laser; *4*—additional mirror

over one metre, the process of measurement will be rendered difficult on account of incoherence of the krypton lamp radiation. These difficulties can be easily overcome by using a gas laser as the radiation source. Since complete coherence is ensured in this case, lengths of hundreds of metres can be measured with a high degree of precision.

The basic diagram of an arrangement for measuring various lengths with the help of a gas laser is shown in Fig. 64.

The laser is composed of mirrors *1* and *2* and a gas discharge tube *3* between them. A light beam emerging from mirror *2* falls on additional mirror *4*, is reflected from it and returns into the resonant cavity of the laser. The phase of the reflected beam depends on the distance the beam has traversed in the course of its travel to the additional mirror and back.

With mirror *4* shifted, the phase and power of the laser vary accordingly. The power variation can be measured by photocell. The periodicity of the variation of the phase and power of the generator depends on the wavelength and equals $\lambda/2$. For example, if mirror *2* is shifted by the distance of 100 wavelengths, then the output power recorded by the measuring instrument (counter) associated with the photocell will change by 200 times. The number of the periods being known, it is easy to find the number of the wavelengths by which mirror *2* was shifted. If the shifting of the mirror was determined by the length to be measured, then the actual measured length will be known with an accuracy of fractions of a micron from the product of the laser operation wavelength and the number of the periods.

Thus, in the National Bureau of Standards (USA) a one-metre long rod was measured by means of a laser interferometer. The length of this one-metre specimen within the measuring accuracy limits ($0.07 \cdot 10^{-6}$) was found to be 1.00000098 m. The result obtained when the length of the same rod was measured by other methods proved to be 1.0000105 m.

There are grounds to expect that in due course the working wavelength of the gas laser will be adopted as the standard of length.

LASERS IN CHEMISTRY

Lasers may prove to be invaluable aids in chemistry. They make it possible to accelerate chemical processes, cause them to proceed with a greater activity and in a definite required direction which heretofore seemed impracticable. Many specialists are of opinion that the application of lasers in chemistry will bring about radical changes similar to those caused by the discovery of atomic energy.

What are the possible uses of lasers in chemistry?

Chemical compounds are known to consist of molecules and the latter, of atoms. Both molecules in compounds and atoms in molecules are bound with one another. The strength of such bondings depends on bonding energy. Atoms in molecules perform vibrations. They vibrate about certain points which correspond to an average energy state of particles. If such a molecule (or a group of molecules) is irradiated with a powerful beam of ordinary light, the amplitude of atomic vibrations in it will be increased in relation to the average state. Since the energy spectrum of ordinary light corresponds to electromagnetic oscillations of most diverse frequencies, the irradiation will increase the amplitude of vibrations of many atoms. In case the irradiation intensity is sufficiently high, some bonds in the molecule may be broken and the molecule will be destroyed. This, however, will be of little benefit. It would be much more interesting if one could succeed in breaking only some definite bonds in the molecule. Then the structure of molecules could be easily changed and reconstructed as required. But how to do it? The omnipotent laser beam is the tool needed here.

If we irradiate a molecule of a chemical compound with a powerful laser beam having one frequency, this frequency will affect only a definite bond in the molecule. The frequency of the laser radiation in this case must correspond to the bonding energy of the molecule. Provided that the radiation intensity is sufficiently high, this particular bond will be broken, while other bonds in the molecule will remain intact.

This specific procedure holds the greatest promise for chemists. By using it, one can selectively break chemical bonds and carry out chemical reactions in a required direction. Obviously, to obtain the required reaction, one should be able to vary the laser frequency. Instead of one variable-frequency laser a set of lasers operating at different frequencies can be employed. Probably, the irradiation of a chemical compound should be performed with several laser beams simultaneously, each having a definite frequency.

As with any new problem, this is certainly associated with difficulties of its own. The frequency required for destroying atomic bonds can be obtained only by using a laser in which the active material consists of the same kind of atoms as those whose bonds are to be destroyed and this is not always feasible. At the same time it is practically impossible to obtain the same frequencies, if the active material of a laser consists of other atoms or molecules: as a rule, different atoms or molecules do not give the same frequencies. Therefore, it is necessary to know the way in which the required frequencies can be obtained. Besides, the laser power should be sufficiently high.

It is hoped, however, that in time these difficulties will be surmounted and chemists will have a powerful tool for creating new chemical compounds.

LASERS IN PHOTOGRAPHY

Photography was invented some hundred years ago and since then no principal changes have taken place in it, except for improvements in the methods of processing and in the quality of photographic materials. Everyone knows how photographs are taken and made. The image of an object of interest is focused on a light-sensitive surface by means of an objective, i.e. by a system of lenses and, as a result, a two-dimensional image of a three-dimensional object is obtained.

Lasers open new, very interesting prospects for photography and offer basically different photographic techniques.

A new method based on the wave front reconstruction became possible due to the use of a coherent light source.

The wave front reconstruction was discovered in 1947 by the British scientist Dennis Gabor. D. Gabor systematically introduced improvements into his method, trying to employ it in electron microscopy. Yet, at that time when no required coherent radiation sources were available, it was very difficult to succeed in the effective utilisation of this method. Emmet Laith and Juris Upaitnieks of the USA revived the original method of D. Gabor. Conditions for the successful realisation of this method were created with the invention of the laser. By now a high-quality three-dimensional (and this fact should be particularly emphasized—three-dimensional!) image of objects has already been obtained. Laith and Upaitnieks in their experiments with a 5 W argon laser were the first to obtain a three-dimensional image of a 0.5 m-long toy locomotive.

The new photographic method requires neither lenses nor objectives.

A laser beam (better a gas laser beam for ensuring maximum monochromaticity) is directed onto an optical system (Fig. 65) which shapes it and makes it wider. The coherence of the beam is not disturbed. A wide laser beam is needed here for covering a broad area within which the object to be photographed is found.

Then the laser beam is directed onto the object. The part of the radiation falling on the object is called the object-bearing beam. The other part of the radiation, having travelled past the object, falls on a mirror. This part of the luminous flux is called the reference beam. The beams reflected both from the object being photographed and from the mirror get onto a photographic plate.

After development such a photographic plate bears an interference pattern. This is the so-called holo-

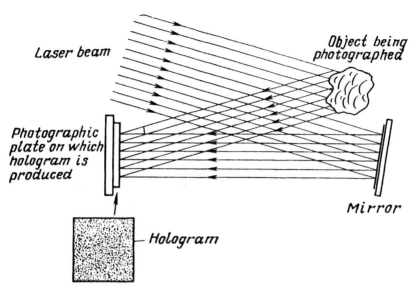

Fig. 65. Making photographs with the help of a laser and wave front reconstruction technique. Reference beam is obtained by means of a mirror

gram. The term "hologram" is derived from the Greek word "holos" which means "whole". The interference pattern is the result of interaction of the waves reflected from the object and those of the reference beam.

As is known, the waves of the reference beam have the same amplitude and length, and are characterised by the same phase relationship. The waves of the beams reflected from the object have different amplitudes and their phases are random. The resulting series of spherical waves each of which originates in a definite point of the surface of the reflecting object are extremely complicated.

This complicated pattern must be recorded on the photographic plate. The wave amplitude is recorded as a darkened area on the light-sensitive layer of the plate. The phase of the wave that has come to the

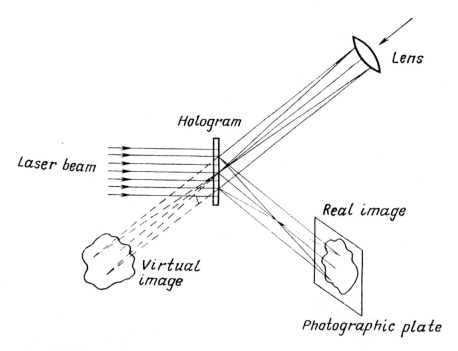

Fig. 66. Reproduction of the image of an object photographed

photographic plate can be recorded by virtue of the reference beam, whose waves, on being added with those reflected from the object, give an interference effect which shows up on the photographic plate as interference bands. In a point where the waves of the reflected and reference beams are in phase, the waves are amplified; in some other point where these waves are in antiphase, they become mutually cancelled. In those points where the phase shift of the waves differs from the two extreme cases, the intensity will be intermediate.

The recording of such an interference pattern gives a hologram. The density with which interference bands are found in the hologram depends on the angle between the direction of propagation of the waves carrying information on the object and the direction of propagation of the reference waves (angle α in Fig. 65).

Thus the entire information which is carried by the waves reflected from the object is recorded on the photographic plate as an interference pattern. But the resulting hologram bears absolutely no resemblance to the object being photographed.

For obtaining an image of the object, the hologram must be illuminated with a laser beam having the same frequency as the reference beam. The scheme of reproducing the image of an object being photographed is shown in Fig. 66. A reference laser beam, on having passed through the hologram, will behave in exactly the same manner as the object-bearing beam did while being reflected from the object, with the photographic plate placed in its path. The process of reconstructing the image when the reference beam passes through the hologram is reverse to that of forming the interference pattern when obtaining the hologram. The similarity of these two processes is the principle on which the wave front reconstruction is based. If we place a photographic plate in the path of propagation of the beams in the plane of image formation, the image of an object will be reproduced on the plate with a very high accuracy. The real image is formed by spherical waves transformed during the reconstruction of divergent waves into convergent ones; at the moment of obtaining the hologram, these divergent waves corresponded to definite points of the object surface. If we look at the hologram from the side of the lens (as indicated by the arrow in Fig. 66), we shall see the virtual image of the object.

The hologram exhibits a number of interesting properties. For instance, it is not at all similar in appearance to the object photographed, so that looking at a hologram you can never guess what image it carries. An exact similarity of the reconstructed and the original waves which were incident on the hologram while it was produced makes it possible to reproduce a three-dimensional image of the object. If we look at the hologram, illuminated by a laser beam, from the side of the arrow in Fig. 66, we shall see a natural three-dimensional picture. It is important to note that this is achieved without recourse to stereophotographic techniques or any other additional devices.

The reconstructed image displays all the features characteristic of three-dimensionality. By changing his posture, an observer will be able to see such details of the object which were hidden before, say, by some other object; turning his head, the observer can look behind this particular object.

If a photographic plate bearing a hologram is broken into pieces, each fragment is capable of reproducing the entire image, though the smaller the fragment, the poorer the image quality will be. This phenomenon is accounted for by the fact that each point of the hologram receives light from all the points of the object being photographed and therefore contains, in an encoded form, the entire information on the object.

One more interesting peculiarity of holograms consists in that several images (up to 150) can be recorded on one hologram, and these images do not absolutely interfere with one another when reproduced.

Holograms with a great number of images are produced by two methods. With one method, objects can be located in different places in front of the photographic plate and illuminated simultaneously with one reference beam. This is a method of coherent superposition, since the light scattered by two objects is coherent and capable of interference.

The other method envisages multiple exposure. The objects of interest are exposed in succession, and each time it is necessary either to change the spatial attitude of the object, or the inclination of the reference beam, or, else, to turn the hologram plate. This method is called incoherent superposition.

Experiments showed the quality of the reproduced images to be satisfactory in either of the cases. By using a divergent light beam during the reproduction (in the cases we have discussed the beam was parallel), the image can be considerably enlarged without recourse to lenses. By illuminating an object with three different monochromatic sources corresponding to three primary colours, a colour three-dimensional image can be obtained.

But not only photography will benefit from the method of producing three-dimensional images. It will find use in high-resolution microscopy, in volumenometry employing stereoscopic and interferometric techniques, for the recording, storage, retrieval and processing of information by optical methods, for the creation of three-dimensional colour cinematography and television.

No doubt that soon a large-auditorium 3-D projector will be constructed on the basis of super-powerful and high-stability lasers. Three-dimensional moving pictures can be made by producing a sequence of holograms on a cine film and then shining coherent light through it and simultaneously brightening up the screen.

The participants of the annual conference in electronics at Stanford University in August of 1965 were the first to see such a performance. Pictures were taken on a standard 35-mm film.

In future, it will evidently be possible to create

three-dimensional television, though this will require a manifold increase in the resolution of the television equipment and, hence, broadening of the television channel frequency band by as much as dozens of times. Laser will come to the aid again—with its help light channels will be feasible having a frequency band of a practically unlimited width.

On getting acquainted with the spectacular possibilities opened up by the use of laser radiation in photography, one can easily imagine those far-reaching advances which television and cinematography will make in the course of time, first of all as regards the creation of three-dimensional colour images. And though the engineering of to-day cannot offer us three-dimensional colour cinema and television systems, such systems will undoubtedly be created in the not distant future.

Holographic principles open the way for optical storage of information. While with the methods employed nowadays information can be stored only in a thin layer of a carrier, the use of holography will allow the information storage in a three-dimensional medium. Coherent light sources will enable a wealth of information to be recorded within a small volume and then reproduced with relatively small distortions.

Experimental tests have confirmed tremendous potentialities of such a principle. The specific density of information recording will reach 10^{12} to 10^{13} bits per cubic centimetre. This means that each cubic centimetre of a crystal can accommodate information contained in a library holding 5 million volumes, 200 pages each, assuming that any printed page contains 1000 words and each word is 7-letter long!

LASERS FOR TREATING OF MATERIALS

The idea of treating materials with the help of a laser was conceived as soon as it was found that a laser beam concentrated on a material of any hardness, be that a copper or steel sheet, a lump of graphite or a crystal of diamond, could cause its instantaneous evaporation. Success achieved with the first experiments in this direction confirmed the possibility of industrial treatment of most diverse materials by using laser techniques. Studies are in progress in this field, and first devices with a laser beam have already appeared in industry.

At present various materials are treated, mainly, by ruby and neodymium glass lasers, since they give maximum output power. As has been mentioned before, the light energy radiated by a laser in a single pulse can reach about a thousand of joules. The pulse lasts for negligible fractions of a second and the power reaches fantastic values amounting to hundreds and thousands of kilowatts.

A pulse lasting one thousandth of a second and having an energy of 0.5 J will be sufficient for burning through a 1 mm-thick steel plate. The hole it will make in the plate will be about 0.1 to 0.2 mm in diameter. Using a beam of the same power, it is possible to weld together two pieces of 0.05 mm-thick foil or two thin wires.

For burning through a steel plate up to 5 mm in thickness, a pulse having an energy of 20 to 100 J is required. In this case the laser beam must be focused onto one point with the help of a system of lenses. Holes made in the metal under the effect of such a beam are, as a rule, of a rather large diameter.

Lasers prove to be particularly effective for treating super-hard materials, such as diamond, corundum and special alloys. Manufacturers of jewels for clockworks and precision instruments, of fine dies, etc., have long been faced with difficulties involved in treating such articles. Thus, drilling of a hole in a diamond die, when performed by conventional techniques, takes more than two hours. The laser drilling machine developed by the Moscow Experimental Research Institute jointly with the Lebedev Physics Institute of the USSR Academy of Sciences performs this operation in less than 0.1 s. Industrial production of such machines for treating super-hard materials in the Soviet Union was started by the "Stankokonstruktsiya" Works. Machines Models K-3 and K-4 can be cited as examples. The K-3 Model is a machine based on a ruby laser. The laser radiation is focused with the help of a high-quality optical system. Light pulses with a duration of 0.5 or 5 ms can be produced every 20 seconds. The pulse power reaches 2 kW and is controlled by a special measuring device. The beam setting accuracy is up to several microns, and the diameter of the focused beam spot can be from decimal fractions of a millimetre to 2 or 3 microns. The machine Model K-3 (Fig. 67) is not large and can be particularly useful for laboratory research, where no high performance characteristics are required.

The machine Model K-4 is of a more perfect design and has a higher capacity. The machine has a water-cooling system and therefore the pulse repetition rate in this machine can be increased to one pulse per second, the pulse duration being 0.5 ms. The machine Model K-4 employs the same optical system as the Model K-3.

It was suggested to use lasers in such operations as balancing parts of precision mechanisms, rapidly rotating magnetic memory disks for computers, etc.

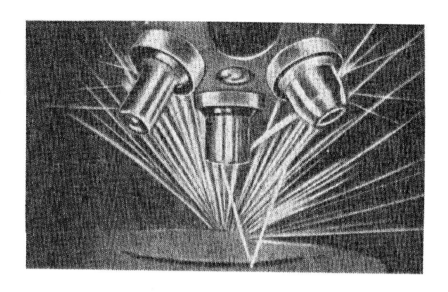

Fig. 67. Laser Model K-3 in operation: laser beam pierces metal

With the help of a laser it is possible to remove excess metal directly from rotating component parts, thus saving the time required for their balancing. Moreover, during such treatment the part involved is not subjected to considerable mechanical loads, as is the case when excess metal is removed by drilling.

In radioelectronics the trend today is toward miniaturisation and microminiaturisation of various units and elements. Many radioengineering devices combine such a great number of diverse functions that if they are constructed by resorting to previous conventional methods, the devices will be both cumbersome and rather heavy. Miniaturisation and microminiaturisation allow the creation of compact units which, while being capable of performing the same functions as their large-sized analogues, yet are hundreds and thousands of times less bulky.

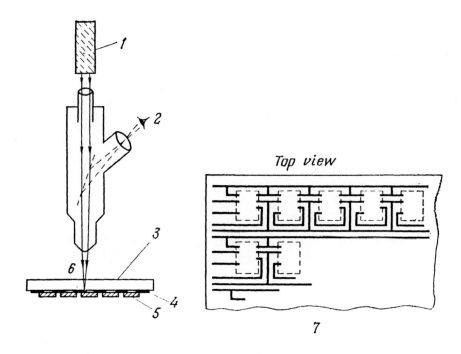

Fig. 68. Connecting units of semiconductor integrated circuits by means of a laser

1—laser; *2*—visual adjustment; *3*—glass support; *4*—deposited metal film; *5*—units of semiconductor integrated circuit; *6*—laser beam; *7*—glass support with deposited wiring circuit

The difficulty arising when assembling such small units resides in connecting their separate elements. A usual electric soldering iron is not the tool to be used here. The omnipotent laser beam comes forward again. With its help one can assemble and connect finest units. The application of lasers will improve both the production techniques and reliability of radioelectronic circuits.

The principle of assembling and connecting the units of semiconductor circuits on an insulation support by means of a laser can be understood from Fig. 68. Leads are prepared by depositing metal on appropriate components of the circuit. The wiring circuit is manufactured on a glass support in the same manner.

The support is placed into a mask, the deposited circuit downwards. Those units of the semiconductor circuit whose tapping contacts are on top are arranged below the support. With the help of a micromanipulator and a microscope they are connected with the corresponding portion of the circuit deposited on the support. Contact between the semiconductor circuit unit and the tapping circuit is ensured by the application of pressure. Then, with the help of the microscope, a laser beam is focused through the glass onto the contact area. The output power of the beam is selected such as to fuse together the metal of the support and that of the circuit unit. On completion of the assembly, the support is placed into a metallic casing.

This method is advantageous in ensuring reliable connections and a high packaging density of the circuit elements.

LASER GYROSCOPES

Gyroscopes are devices widely used as instruments in the navigation systems of ships, in automatic flight control systems of aircraft and space vehicles, etc. The basic component of the gyroscope is a small heavy wheel rotating at a high speed. The gyroscope can retain the direction preset to its spin axis and resist any changes of this direction caused by disturbing forces acting on the gyroscope. The gyroscope stability is the higher the higher the rotation speed of its wheel, the latter amounting to 30 000 r.p.m. and over.

Mechanical gyroscopes, however, are vulnerable just on account of the presence of rotating parts in

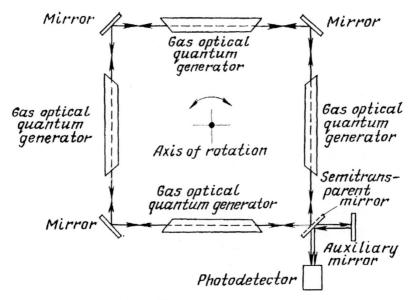

Fig. 69. Laser gyroscope

them, since this impairs their reliability. Besides, the sensitivity of such gyroscopes is not always sufficient. At the same time reliability of equipment functioning in flying vehicles is, certainly, of crucial importance.

Latest researches have shown that lasers can be employed for constructing such a navigation instrument as the gyroscope.

The laser gyroscope is an instrument of a new type operating on the principle of the well known Doppler effect. This gyroscope has no moving parts and therefore theoretically it must combine such advantageous properties as long service life, high sensitivity and stability. The output signal of the gyroscope can be presented in a digital form and therefore the laser gyroscope can be conveniently employed in combination with an electronic computer.

A laser gyroscope is a system of four He-Ne gas

lasers arranged so as to make a square (Fig. 69). Set at each corner of this square at 45° to the axis of the lasers is a mirror, which ensures the circulation of the radiation of the lasers along an annular path. Since each of the four lasers emits from both ends of its tube, two light beams are created in the gyroscope, that are moving along a circle in opposite directions. In case the laser square remains stationary, the both beams traverse equal distances. But if the lasers are mounted on a platform rotatable about an axis perpendicular to the plane of the four lasers, then the beam travelling in one direction will have to cover a somewhat greater distance to reach its initial emergence point than the beam travelling in the opposite direction. The result will be a frequency shift (Doppler effect). This shift can be measured by optical methods. To this end, one of the four mirrors located at the corner of the square is made partially transmitting, so that a small portion of the light of the oppositely moving beams could pass through the mirror and the rest of the light should be reflected and continue its circulation.

By placing an additional mirror in the path of one of the beams emerging from the partially reflecting mirror, at right angles to the direction of the beam travel, the beam will be caused to be reflected from this additional mirror back to the mirror at the corner of the laser system and the both beams will emerge from the system in the same direction. These beams are then intercepted by a photodetector and their mutual shift results in the origination of beats at an audio frequency equal to twice the Doppler shift.

The angular velocity can be found from the formula

$$\omega = \frac{\Delta f p \lambda}{4A}$$

where ω is the angular velocity; Δf is the frequency shift (Doppler shift); λ is the laser radiation wavelength; p is the perimeter of the square; and A is the area of the square.

In an experimental model of a laser gyroscope the optical length of its arm was about 1 m with the radiation wavelength of He-Ne lasers equal to 1.153 μ. The frequency difference signal obtained at the photodetector output was 250 Hz per degree of rotation per minute. The frequency of the output signal obtained at the rotation velocity of 2 deg/min was 500 Hz and at that of 600 deg/min, 150 kHz.

The laser gyroscope of such a design has large dimensions and this is a disadvantage. These dimensions can be essentially diminished by using semiconductor lasers instead of gas ones. Though a shorter path of the beam tells on the gyroscope sensitivity, this undesirable effect could be compensated for by employing lasers operating at shorter wavelengths (on the order of 0.71 to 0.84 μ).

Though the dimensions of gyroscopes on gas lasers are considerable, but despite this fact they can be successfully used even now, e.g. in ships, where no particular limitations are imposed on the weight and dimensions of the gyroscope.

In one experimental model of a laser gyroscope an equilateral triangular resonator was employed with corner mirrors spaced at 138.56 cm. The gas laser radiation wavelength was 6328 Å. Experts are of opinion that with this design the adjustment of the instrument can be facilitated and optical aberrations minimised.

A gyroscope of a similar design but with different dimensions of the resonator, with the distance from the centre of the triangle to the corner of 10 cm, is capable of measuring angular velocities less than

0.001 deg per hour. It can be employed as a very precise standard of angular position. The resolution of the instrument is less than 0.25" and drift, less than 5" a day.

At present work is in progress for the development of small-size laser gyroscopes. It is contemplated to build a gas laser gyroscope having a weight less than 0.9 kg and overall dimensions of 0.5 cu dm.

LASERS IN DETECTION AND RANGING

Radar systems use transmitted and reflected electromagnetic radiation for detecting various objects and determining their spatial coordinates. Such objects can be found on the surface of the earth, in air or at sea. They can be stationary or in motion; in the latter case certain parameters of their motion are determined. Devices that solve such a problem are called radars.

Until recently electromagnetic waves belonging to the radio range (metre, decimetre and millimetre ones) were employed in radar systems. These systems came into being during the World War II. In the post-war period the radar techniques found rather extensive applications and reached high scientific and technical standards.

Laser radar systems constitute one of new branches of modern quantum electronics. Considerable advances made in this field during a comparatively short period became possible due to numerous theoretical and experimental research carried out by scientists in the Soviet Union and other countries.

Laser radar techniques are based on the use of the optical range of electromagnetic oscillations generated by lasers. Light has long been used as a means for detection and observation. Before the advent of lasers search lights were the main apparatus employed for detecting, observing and tracking of targets. The energy radiated by such light sources was used only to illuminate the object, the observation being performed visually. Naturally, the possibilities of search lights were limited.

With the appearance of coherent light sources the methods of using light energy in observation facilities have changed. Lasers made possible the formation of light pulses having a small duration and an energy level sufficient for recording a signal reflected from an object at the receiving station.

The use of the optical range in laser radar systems is justified for many reasons. Thus, as compared with conventional radars, radars operating in the optical range have a higher directivity of radiation with comparatively small dimensions of their aerial devices, a better resolution with regard to angular coordinates and range. An optical radar is capable of detecting an object within a pre-set coverage using a transmitter whose power is millions of times less than that of transmitters of radar stations operating in the millimetre range.

Laser radars are almost insensitive to the effect of intentional interferences. These interferences can affect a laser radar only when their source is within the radar beam. But taking into account that the beam is very narrow, the probability of an interference source being found within the radar beam is very small. The effect of interferences can also be reduced by using special filters. Laser radars are noted for their small overall dimensions and weight.

While featuring the above-mentioned advantages, laser radars are not free from certain disadvantages. Among these considerable attenuation of laser radiation in case of fog, rain and snow should be pointed out, this factor imposing a limitation on the distance over which laser radars are effective. In this connection laser radars appear to be most promising for operation in the outer space where there is no atmosphere, and in mountains, above the level of precipitation.

The block diagram of a laser radar resembles that of a conventional radar. The laser radar determines the same characteristics of objects: distance, altitude, angular coordinates and velocity. The diagram of a "Colidar" laser radar is shown in Fig. 70. The "Colidar" system consists of three main parts: a transmitting device, a receiving device and a data processing device. The transmitting device comprises a ruby laser, a collimator serving to narrow the laser beam, and an optical shutter which shapes a rectangular light pulse. The receiving device consists of a concave mirror by means of which the light reflected from an object is collected, a narrow-band filter which diminishes the background noises or interferences, a photomultiplier which converts light oscillations into electrical ones, and an amplifier of electrical oscillations. The data processing device gives the coordinates of the object involved.

Distance to the target is found from the time period between the sending of an emitted (sounding) pulse and the incoming of the reflected signal. The "Colidar" laser radar emits a light pulse which lasts for 0.003 s and has an energy of 2 kW. The weight of this radar is 11 to 14 kg and coverage under normal meteorological conditions, 15 to 30 km. It can discriminate between two objects that are at a distance of 10 km from the radar and spaced 3 m apart.

Other types of lasers, e.g. gas lasers can be used as radiation sources as well.

According to publications, a laser radar with a mir-

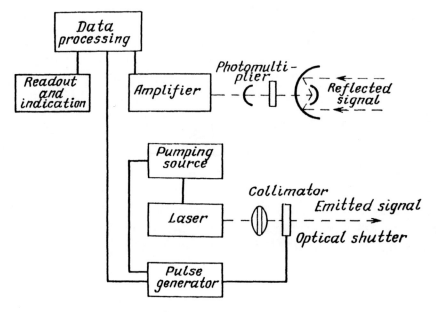

Fig. 70. Block diagram of a laser radar

ror having 60 cm in diameter and an average radiation power of 70 W can measure the distance to a rocket having a diameter of 6 m with an accuracy of 1.6 km when this distance amounts to 160 000 km. Applicability of laser radars for underwater operation is under investigation. Good results can be expected here with the use of lasers radiating in the blue to green region, since sea water is most transparent within this region. A laser underwater radar intended for the detection of submarines, torpedoes and mines is expected to have a coverage of several kilometres.

The velocity of an object is determined by using the well-known Doppler effect.

LASER RANGE FINDERS

In the last few years an urgent need has been felt for the creation of an instrument which would be fit for measuring distances to various objects under various conditions with a high accuracy and, at the same time, be free from the disadvantages inherent in radar and optical range finders. The main disadvantage of a radar is its broad aerial directivity pattern, which causes interferences from neighbouring objects. Besides, the radiation of a radar can easily be detected. For distance measurements to be made with a high accuracy by using an optical range finder, this instrument must have a very large base.

A laser range finder with its powerful signals and monochromatic thin beam is free from all these disadvantages. The operation principle of such a range finder is analogous to that of a conventional radar, though, naturally, there are some specific features in its design.

Shown in Fig. 71 is a block diagram of a laser range finder. How does such a range finder operate?

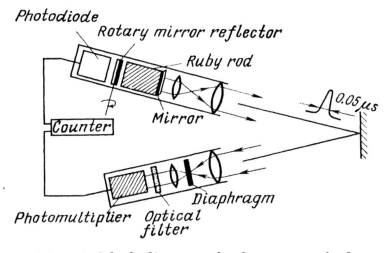

Fig. 71. Block diagram of a laser range finder

Fig. 72. Laser range finder

A laser beam is directed to an object through a transmitting objective tube and, after falling on the surface of the object, is reflected from it. Part of the reflected signal is received by a receiving objective tube which is provided with a narrow-band optical filter at the outlet. This filter makes possible the separation of the reflected signal even against the solar radiation background. Then the signal comes to the input of a photomultiplier. An amplified signal triggers a generator of gating pulses. From the number of the pulses that have arrived at the input of the instrument during a certain period of time the distance to the object can be determined.

Structurally a laser range finder consists of two parts: a head and a power supply unit. The head incorporates a sighting device, a receiver and a transmitter. The transmitter incorporates a ruby laser, pumped with the help of a xenon tube. The head also accommodates an indicator, a computing device, and a photomultiplier. The general view of a laser range finder is shown in Fig. 72.

The coverage of this apparatus is up to 6 km and under the conditions of good visibility, to 10 km. Its measuring accuracy is 10 m irrespective of the dist-

Fig. 73. Soviet laser range finder Model ГД-314

ance being measured. The apparatus is easy to operate: the operator must only select an object and press the control button. The distance to the object will be instantaneously displayed as digits in the indicator window.

Reports have been made of experiments made with an appropriately modified laser range finder functioning as an altimeter. It can also be used for precise calibration and checking of conventional altimeters under various flight conditions and various character of the Earth's surface.

Experimental flights showed that with the help of a specially constructed airborne laser altimeter the height of up to 300 m could be measured with an accuracy of up to 1.5 m. It is interesting to point out that when the surface of the Earth is covered by a forest, the reflected signal has the character of a "double echo". This phenomenon is conditioned by the reflection from the tops of the trees and from the surface of the Earth. The double echo effect allows measuring the absolute height of the trees and makes laser altimeters applicable under such conditions where any conventional radar altimeter would be useless.

Figure 73 shows one of the laser range finders made in the Soviet Union—Model ГД-314. The apparatus is intended for precise measuring of distances within a range of up to 2000 m. The measuring accuracy within the entire range is 2 cm. The radiation source is a semiconductor diode laser with the wavelength of 8600 Å. The radiation power is 0.5 mW. The range finder consists of an optical transceiver unit weighing 6 kg, a measuring package weighing 5 kg and a power supply unit weighing 15 kg. Storage batteries ensure continuous operation of the range finder during 50 hours.

LASER TRACKING OF SATELLITES

Numerous artificial satellites are now ploughing the circumterrestrial space. These satellites serve most diverse purposes. Geodetic satellites, for example, help to obtain more exact information about the configuration of the Earth.

The velocity and altitude of the artificial satellites are not constant, but vary with time. These characteristics are very important for acquiring more precise data concerning the parameters of our planet. For instance, the processing of the data collected during the flights of satellites allowed a more complete specification of the depth in one of the regions of the Indian Ocean.

The application of lasers in space technology is one of the main tasks set before numerous companies and research centres, the Air Force and other armed services of the USA.

At present tracking of artificial Earth satellites with the help of passive optical means can be practised only in those cases when the satellite is illuminated by the Sun and the ground tracking stations are in the shade. For obtaining all the necessary information, e.g. on the distribution of the gravitational field of the Earth, one should know the parameters of the entire orbit of the satellite and not only such data as can be collected while the satellite is illuminated by the Sun. The use of lasers allows the information on the orbit parameters to be obtained both in the daytime and at night, the measuring range being thus substantially broadened. An experiment carried out in the United States furnished an answer to some of the problems associated with the application of lasers for tracking flying vehicles in the outer space (Fig. 74).

An optical telescope sighted at the space vehicle carried a laser which periodically illuminated the satellite. A corner reflector array located on the satellite reflected the laser radiation towards the tracking system. The reflected radiation was received by a photomultiplier. The distance to the satellite and the tracking error were determined from the detected signal.

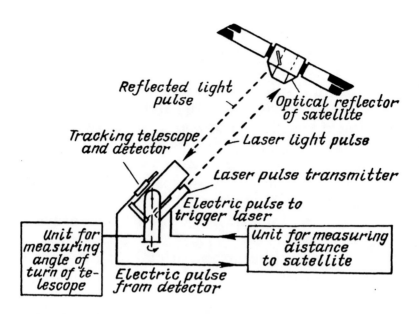

Fig. 74. System for tracking artificial satellites

The experiment was carried out with the S-66 satellite launched into an orbit at altitude of 1000 km. The satellite was equipped with a magnetic stabilization system designed to maintain it in such an attitude that the axis of the satellite should always be directed parallel to the magnetic field of the Earth. The satellite slowly rotated about its axis. The corner reflector array on the surface of the satellite was directed towards the Earth when the satellite was within the north hemisphere.

The reflector array was assembled from 360 mirror corners, 2.6 cm in cross section each. The radiation from the artificial Earth satellite was reflected at a very small angle (within 10^{-4} radian, Fig. 75).

Presented in Fig. 76 is a diagram of a transmitter, including a ruby laser and collimator optics, ensuring the angular divergence of the beam on the order of 1 milliradian. An eyepiece and a pentaprism allow

Fig. 75. Satellite-borne reflector array

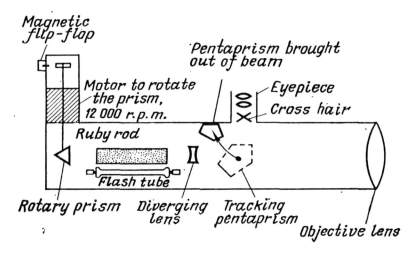

Fig. 76. Diagram of a laser transmitter

the operator to perform accurate tracking of the satellite. When the laser is triggered, the prism is automatically brought out from the beam. Another reflecting prism is rotated by a motor with a speed of 12 000 r.p.m. and ensures modulation. The pulse repetition rate of the laser is 1 pulse per second and its output energy is about 1 J.

The receiving device which is usually employed in

Fig. 77. Ballistic camera for tracking artificial Earth satellite

tracking telescopes is located in place of the cine camera.

A 47-cm telescope of such type at the Wallops Test Station (Wallops Island, USA) was used in the experiments described (Fig. 77).

The principal diagram for determining the distance to an artificial Earth satellite is shown in Fig. 78. The sounding pulse of the transmitter is used for triggering, scanning and indication of the reference pulse on the screen. The pulse reflected from the satellite produces a luminous mark on the indicator screen.

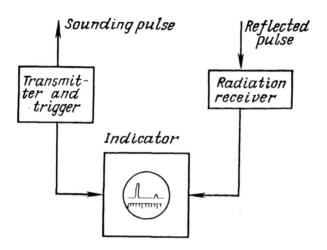

Fig. 78. Diagram of a system for determining the distance to an artificial Earth satellite

The distance between the pulse marks on the screen indicates the delay of the reflected pulse in relation to the reference one. The intensity of the reflected signal is sufficient for its being detected by a photodetector at night. It is believed that a photographic picture can also be recorded as reflected flashes against the starry background, provided that the pulse power is increased and a large ballistic camera is employed. If so, accurate angular tracking will be feasible in twilight and in the shade of the Earth.

Photographs cannot be taken in the daytime, but with the help of an electron photodetector and by using special narrow-band filters the signal can be separated against the sky background.

The system operation range is rated to be up to 1500 km, the vehicle flight coordinates being measured every 10 s. Theoretical calculations suggest that the results thus obtained allow the satellite trajectory to be determined with an accuracy of up to 30 m.

The main difficulty in this experiment is to preclude the beam deviation from the target. But the use of electronic computers will, probably, offer a solution of this problem as well.

American scientists created several such systems which were located in various places of the globe. Experiments with the tracking of the artificial Earth satellite "Explorer-22" were conducted in 1964.

But at first the attempt at receiving the reflected signal ended in failure. Scientists attributed it to difficulties involved in accurate tracking of the satellite and to atmospheric interferences.

French scientists proved to be more lucky and succeeded where the Americans had failed. In the end of January 1965 a series of experiments were conducted by the staff of the Saint Michel-de-Province Observatory on tracking the same satellite with the help of their own ground equipment. The duration of the sounding pulse was $3 \cdot 10^{-8}$ s. The distance to the satellite was 1517.99 km and it was determined with an accuracy of up to 8 m.

When estimating this experiment, one should take into account a tremendous distance to the satellite (more than 1500 km), its cosmic speed ($2 \cdot 10^4$ km/hr), small dimensions (the diameter of the "Explorer-22" being 60 cm) and short duration of the pulse, which was only 3 hundred millionths of a second! One French astronomer made a witty remark, saying that this experiment could be compared with an expert shot at the eye of a fly darting at a speed of 100 kilometres per hour from the distance of 5 kilometres.

Somewhat later, in February 1965, the same experiment was successfully performed by American scientists, when the satellite was flying at the altitude of 950 km. The reflected beam was received by the ground equipment of the Hensfield Air Base.

The success with this experiment allows one to come to the conclusion that with the help of a laser it is possible to measure distances between remote points of the earth surface and send various signals to artificial Earth satellites.

LASERS IN SPACE EQUIPMENT

During the period prior to the "Gemini-7" spaceship flight with two astronauts on board, the development and tests of a laser air-ground communication system have been completed in the United States. This system envisages the communication of the spaceship with the ground station by means of a laser beam. The given system was supposed to be employed when launching spaceships in accordance with the "Gemini" and "Apollo" programs, for the transmission of speech and communication through the plasma screen during the landing. Besides, according to the opinion of specialists, with this system it is possible to determine the distinctive characteristics of the coherent beam passage through the atmosphere and the potentialities of the space equipment for determining the location of the ground communication station.

The system was tested with a jet aeroplane. The laser communication system consisted of three parts: a ground laser beacon, an airborne laser transmitter, and a ground receiver. The laser beacon served for aiming the airborne transmitter at the ground receiving station (Fig. 79). For ensuring a continuous tracking of the laser transmitter by the receiver, the laser transmitter and the beacon were mounted on the platform of a conventional tracking device slaved with the near-by aerial of the radar station.

During the tests of the laser communication system the jet aeroplane performed regular daily flights, usu-

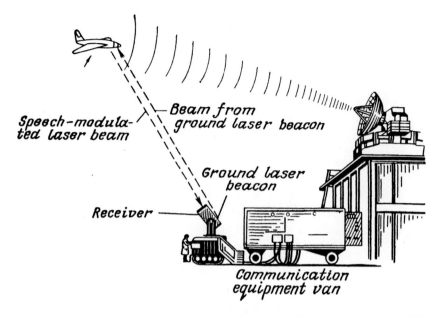

Fig. 79. Laser communication with an artificial Earth satellite

ally just before the sunset, flying at a height of 3 to 12 km above the ground station in an anticlockwise direction at a speed of 500 to 1300 km/hr describing elliptical paths totalling about 1100 km, so that the slant range to the receiving station was 8 to 24 km.

Since the laser communication system was ultimately intended for use in a manned orbital spacecraft, the flights of the jet aeroplane were such as to simulate the tracking angles and flying speeds of a spacecraft when in the near-earth orbit. One of the problems to be solved during the tests of the system was to find out whether the astronaut would be able to detect the laser beacon and then aim the beam of his transmitter at it with an accuracy sufficient for establishing a reliable one-way telephone communication.

Another objective pursued during the tests was to evaluate the quality of speech signals transmitted by the infra-red beam, the sequence of unfiltered pulses and background noises. The American specialists hoped that the results of these tests would help to reveal the influence of the terrestrial atmosphere on the propagation of optical oscillations.

The signals received by the ground station in the course of the first tests proved to be unintelligible. The tests were carried out again and the recordings obtained could be understood, though their quality was poor. Nevertheless, the experiments showed the possibility of employing lasers for work in the air-ground communication system.

The experts engaged in this project came to the conclusion that the creation of an operable communication system of such kind calls for the development of a more reliable aiming and stabilization system.

A simplified block diagram of a laser communication system is shown in Fig. 80. A beacon which is a gallium arsenide semiconductor laser functions as a reference point for the satellite-borne laser transmitter to be aimed at the receiving station. Since the coverage area of the transmitter beam at the reception point for the distance of 16 km is only 30 m in diameter, the aiming of the beam was of great importance. The divergence of the beacon radiation was 10^{-3} radian; the pulse power was 1 W and its duration, 5 µs. The beacon operation wavelength was 8400 Å. The sighting telescope of the transmitter employed in the flight tests of the laser communication system was provided with an image converter sensitive only to the narrow-band emission of the beacon. The receiver employed a collector having 75 cm in diameter and consisting of a non-spherical primary reflector and a spherical secondary mirror. Background noises were cut off by passing the received signals through a multilayer filter with a pass band of 50 Å. The filtered

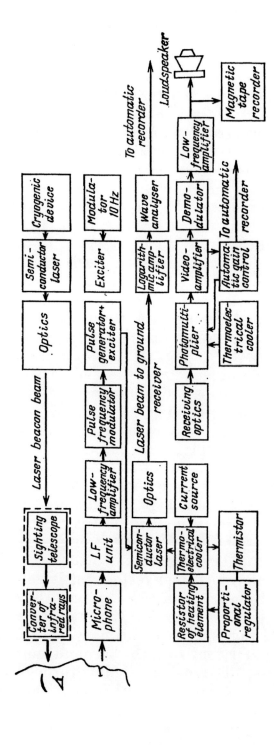

Fig. 80. Simplified diagram of a laser communication system with artificial Earth satellites

signal was passed through a diaphragm that restricted the viewing angle to 0.001 radian. After that the light energy was focused onto a photomultiplier, and its output signal was sent to a video amplifier.

To bring the noise level below the operation threshold of the demodulator in the receiver, the noise voltage of the video amplifier was passed through a bandpass filter, rectified, and then used for adjusting the photomultiplier gain. A multivibrator produced a train of standard output pulses whose frequency was determined by that of the input video pulses. The narrow-band filter ensured demodulation, and the processed signal was then amplified and sent to a tape recorder and a loudspeaker.

The receiver was protected against casual direct sunlight by a special automatically actuated shutter mounted on the collector. This shutter opened only when the solar radiation level was within safety limits.

The 4.5-kg laser transmitter was equipped with a sighting telescope and incorporated a GaAs semiconductor laser, modulation and control equipment and storage batteries. The power of the transmitter output pulse was 5 W.

During the transmission the operator used a microphone built into the transmitter. The speech signal came to an amplifier where amplification and compression of the dynamic range were performed as required for protection against interference. The low-frequency channel width was 0.3 to 3.0 kHz.

The amplifier output signal was sent to a pulse frequency modulator. This modulator produced a train of pulses which were passed to the exciter of the semiconductor laser—a delay line and a transistor switch. When the transistor switch was actuated by the output voltage of the modulator, the delay line was discharged through a pulse transformer, thus exciting the laser which emitted at 8900 Å. The laser temperature was maintained at a definite level (16°C) by means of a thermoelectrical cooler. The infra-red beam emergent from the laser was directed to the optical system through a lens with a focal distance of 7 cm. The divergence of the laser beam was $2 \cdot 10^{-3}$ radian.

As reported, during the flight of the "Gemini-7" spaceship, the astronauts, after having made two unsuccessful attempts to communicate with the Earth, the failure being on account of some troubles in the ground equipment, established the communication by the laser beam when making the 105th orbit. The communication was satisfactory, lasted for 2 minutes, and some information was transmitted during it.

Thus, the feasibility of laser communications with space vehicles was proved.

COMMUNICATION WITH SPACECRAFT DURING ATMOSPHERIC RE-ENTRY

For space flights to be successful, it is a prerequisite that the ground control over the spacecraft flight should be continuous. As is known, during the earth re-entry, the spacecraft body becomes strongly heated, on account of a high velocity with which it travels and a tremendous resistance offered to it by the dense atmosphere. The gas in the boundary layer between the shock wave front and the spacecraft surface becomes ionised. A high-temperature plasma is thus formed around the space vehicle, which may fully envelop it and the high-frequency radio communication units it carries. This moment is critical for the radio communication. The concentration of free electrons in the layer of ionised gas is very high, and this makes such layer conducting. Radio waves are either reflected from the plasma layer or absorbed by the ionised gases. During the break in the radio communication with the spaceship it can be no longer program-controlled. Some important information, in case the space vehicle gets in trouble while returning, may be lost. The maintaining of constant radio communication with the returning space vehicle for information exchange is still more necessary if such a vehicle is manned, for it is quite probable that during this most difficult and responsible period they may need help from the Earth.

The plasma layer enveloping the shell of a missile disturbs the operation of its antimissile and antisatellite systems which can function only on continuously receiving the information about the presence of such targets and their location.

In view of these and many other reasons the problems of ensuring communication with space vehicles during their atmospheric re-entry at supersonic velocities and re-establishing such communication when it is disturbed by the exit gases of the rocket engine have long been most urgent. Special measures are reported to be developed for ensuring radio communications through a plasma layer. One of such measures is the determination of an optimum frequency range within which plasma features maximum transmittancy. Another measure can be the selection of an appropriate configuration of a space vehicle, so that in certain places the thickness of the plasma layer would be small and the attenuation of communication signals passing through it, insignificant. For diminishing the thickness of plasma envelope, it is, probably, advisable to make the head portion of the space vehicle pointed. Space vehicle aerials should be mounted as far as possible from its head portion, i.e. in such places where the plasma envelope can be "washed off". It is suggested to set up special magnetic fields near the space vehicle aerials and thus create better conditions for the passage of radio waves. Such a project, however, adds to the weight of the vehicle.

The concentration of free ions in plasma may be reduced by introducing a special substance into the plasma flow near the head portion of the space vehicle; this substance will lower the temperature of the gas and cause recombination of electrons and ions. This can be achieved, however, only in case the space vehicle travels at a comparatively low speed. The substance should be introduced in the vicinity of the aerials so as to neutralise positive ions in these places. The substance was suggested to be introduced in the form of negatively charged microscopic droplets. Finally, it was proposed to cool the outside surface of the vehicle with water jets for creating plasma-free areas near the aerials.

All these measures which, by the way, are not at all easy to realise and most of which lead to a considerable increase in the weight of the space vehicle, cannot guarantee stable communication at the moment of its atmospheric re-entry.

The problem of ensuring stable communication with the space vehicle also arises when exploring our neighbour-planets Mars, Venus and Jupiter. These planets are known to have atmospheric envelopes of a certain density. Unless special measures are taken, the communication with such spacecraft will be broken as soon as they enter the atmosphere of these planets.

This was the case with an American space probe launched in 1964 and missioned to explore the composition and density of the Martian atmosphere. The communication with the space probe stopped as it entered the dense layers of the Martian atmosphere, and no information on the composition and density of the latter was obtained.

As is known, a laser beam can rather easily pass through plasma. In view of this fact, and seeking for an effective solution of the communication problem, extensive investigations have been carried out for studying the conditions of the modulated laser beam passage through plasma. It was established that during the most critical part of the flight, namely, during the period of atmospheric entry (or re-entry) the communication with space vehicles can be ensured with the aid of a laser. Certain and rather essential difficulties are encountered here as well. They are: the choice of an effective modulation of the beam, elimination of the atmospheric absorption effect, accurate tracking and holding of the beam at the point of the vehicle location, etc.

Nevertheless, it is quite evident that these difficulties will be overcome and the problem of ensuring reliable communication with space vehicles during their entire flight will be successfully solved by the use of lasers.

DETECTION AND COMMUNICATIONS UNDER THE SEA

The main communication means of modern submarines are radio communication facilities. Some scientists abroad are of opinion that only long-wave and superlong-wave bands are fit for communications with submarines. It is known that long radio waves can not only round the Earth but also penetrate deep into the sea. No communication with an object found at a depth can be effected on medium, short, or ultrashort waves. The waves belonging to these bands are almost completely scattered or absorbed by the water medium and do not penetrate deep into it.

To realise a long-wave communication is not an easy matter either. Powerful transmitters will be required for this. To give an example, it will suffice to mention that one of special radio broadcasting stations operating on superlong waves uses a transmitter whose power is 2000 kW. The aerials of this station are mounted on 26 towers up to 300 m in height. Naturally, no such system of aerials can be located on a submarine and hence submarines employ only one-way communication systems, for reception. For transmitting any message, the crew must surface the submarine and send the message using a different frequency range. Meanwhile the submarine bearings can be easily taken and the submarine will be thus detected. The secrecy of information transmission is ensured by using special high-speed devices capable of transmitting a 200-word message per second.

There is still another problem: to ensure communication between neighbouring submarines, ships and aircraft. Here again two-way communications are effected using short-wave bands and this is a serious disadvantage from the standpoint of secrecy and security.

As a way out of this difficulty it is suggested to employ lasers for communications under the sea. In the USA it is planned to develop a laser communication system operating on the periscope-to-periscope principle. In case of success such a laser system is expected to be most high-speed and noise-resistant, as well as adequate for meeting information transmission secrecy requirements.

An extensive research program sponsored by the US Navy Department was started in the United States on the use of lasers and for the development of new apparatus. The objective pursued by the program was twofold: to study the optical radiation transmission conditions in sea water and develop new models of detection and communication equipment.

The investigations revealed that the radiation propagation range in sea water depends, mainly, upon the absorption of radiation by the substances dissolved in water and upon the scattering of radiation on particles suspended in the water. In some water samples absorption was found to be predominant and in others, scattering. At the same time it was established that water, similarly to the atmosphere, possesses different spectral transparency. Radiation in the red region of the spectrum (corresponding to that of a ruby laser) was shown to be absorbed by sea water stronger than radiation in the blue-green region of the spectrum. Therefore, the latter radiation can propagate in sea water over considerable distances and, hence, frequencies belonging to the blue-green region of the spectrum should be preferred for the detection and communications under the sea.

A powerful source of coherent light was developed by the Laser Advanced Development Center (USA). This source consists of an optical quantum generator and a potassium dihydrogen phosphate or an ammonium dihydrogen phosphate crystal displaying non-linear characteristics.

A light beam generated by the laser is directed onto this crystal which separates the higher harmonic. The radiation takes place at 5300 Å, which corresponds to the green region of the spectrum with the spectral line width of about 2 Å. The beam divergence is as small as 1 milliradian, and the radiation power is 10 kW. Radiation with the wave length of 2896 Å corresponding to the ultra-violet region of the spectrum was obtained with an $Ar-CO_2$ gas laser. Radiation at the wave of 3125 Å, corresponding also to the ultra-violet region of the spectrum, was obtained with a laser employing gadolinium-doped silica glass as the active material and excited by a xenon flash tube.

A model of an underwater optoelectronic radar built around a laser was also developed in the United States. This apparatus employs a conventional image-scanning circuit, and consists of a transmitting, a receiving and a recording system. The radiation source in the transmitting system is a laser; its scanning beam irradiates a definite field of view. The receiving system incorporates a narrow-band optical system operating in synchronism with the laser beam. The radiation reflected from the object is received by a photomultiplier. The recording system of the radar produces an image of the object involved.

The resolution of such an arrangement is higher than that of present-day underwater television equipment; the operation range of the arrangement is several kilometres, whereas the present-day underwater television equipment coverage is about 140 m.

It is also contemplated to employ lasers for detecting submerged mines without running the risk of triggering their fuses, sensitive to the operation of sonars. It is considered that laser beams can be used for homing torpedoes and other unmanned underwater vehicles.

OTHER MILITARY APPLICATIONS OF LASERS

Besides those possible military uses of lasers which we have already mentioned, there are other fields of their application that appear no less important in the opinion of foreign specialists. Many American companies sponsored by the US Department of Defence are engaged in elaborating new equipment with the use of lasers for military purposes. The laser technique is also rapidly developing in other economically advanced capitalist countries such as France, Great Britain, the FRG and Japan.

Assuming that high-power lasers can be created in the near future, American specialists advance projects of using them as a weapon in anti-missile and anti-aircraft defence systems.

The main advantages offered by the laser weapon in this case are: a high propagation velocity of radiation, equal to the propagation velocity of light and some tens of thousands of times exceeding the speed of antimissiles; absence of scattering and losses in the medium when this weapon is used in the outer space; less sophisticated ground equipment. The application of the laser weapon for destroying the atomic and thermonuclear warheads of ballistic missiles will cause a much less radioactive contamination of the atmosphere and the outer space than the use of antimissiles with nuclear warheads.

To destroy an enemy missile, not to let it reach the target, it is sufficient to put its control system out of action. This can be done by burning through the missile shell or rudders by a laser beam. This will cause vibrations in the missile and result in its complete destruction.

Figure 81 shows a block diagram of an anti-missile system based on the use of lasers. Such a system must have a receiving unit for processing the signals incoming from the early-warning and target-tracking radar stations. These signals contain information on the coordinates of the approaching missile. The tracking station must aim at the target an optical radar in which a laser serves only for determining the distance to this missile.

Such an optical radar can furnish very precise data on the coordinates of the target, and these data are

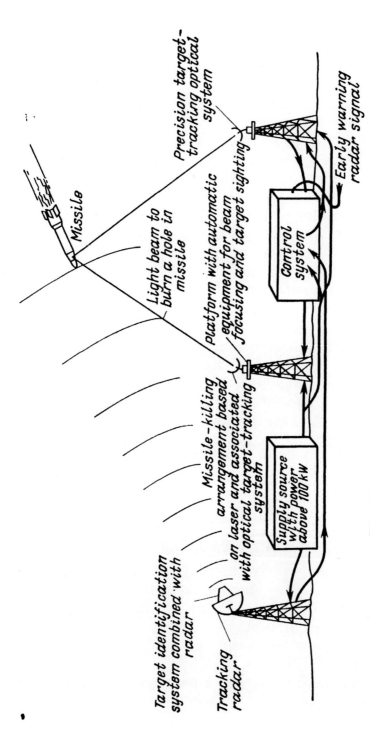

Fig. 81. Diagram of an anti-missile system

used to actuate another system employing a high-power laser, designed for destroying the target. The optical radar will focus and aim a powerful laser beam at the most vulnerable point of the missile during a period of time required for a hole to be burnt through in the missile.

The authors of this project believe that the use of the laser weapon would make unnecessary the identification of decoy and combatant enemy missiles, since the laser weapon ensures rapid destroying of both combatant and decoy missiles.

Another possible anti-missile laser defence system is a project of an orbital space station equipped with target-detecting and tracking radars, as well as with lasers which can be excited by the solar energy. One of the main difficulties associated with this project is the provision of a platform capable of ensuring sufficient stability of the equipment and accuracy of the target tracking. For the target to be destroyed, a laser beam must be directed onto it during a long period of time, and this will require the use of high-speed servounits. Serious difficulties are also encountered with the methods of focusing onto targets of energy having a sufficient density. Therefore, though lasers have already been created featuring an increased pulse power amounting to several tens and even thousands of megawatts, the creation of a laser weapon that could be applicable in anti-missile and anti-aircraft defence system still requires the solution of a number of complicated engineering problems.

Much attention is paid to the development of special phased gratings which make the radiations of several lasers to be collected strictly in-phase into one common beam, with a view to increasing the radiated power.

It was suggested to use lasers for semiactive guidance. The essence of this method is as follows. The beam of a laser which is not locked with an object to be guided, say, a projectile, is aimed at the target. After the projectile has been launched, the operator, with the help of a programmed device, gathers it into the beam and holds the projectile in the beam during the flight till it reaches the target. After the projectile has hit the target, the laser is used for guiding to other targets.

Such a system is supposed to be applicable for guiding anti-tank projectiles. The main advantage of this guidance system is its being jamproof. A strictly directive narrow laser beam always ensures a high accuracy of guidance.

Special devices for guiding aerial bombs are reported to have been developed for the US Air Force. The guidance is effected by a laser radiation reflected from the target, the radiation which illuminates the target being furnished by other independent laser sources. Laser-guided bombs are conventional aerial bombs, only instead of ordinary fin assemblies they are provided with assemblies controlled by a unit with a laser homing head. Tests of these bombs showed that the probable radial error of striking the target is about one tenth that observed when using conventional bombs. The Americans used laser-guided bombs in their air raids in Viet-Nam.

Laser devices are coming into use in surveillance and reconnaissance systems. The first laser cameras for aerial reconnaissance borne by some craft of the US Air Force allow the photographs to be taken during night flights.

American military experts consider that laser weapons can be used against enemy manpower. Possible types of such a weapon may be a laser pistol, a laser rifle, or a laser gun. For making a laser pistol, for

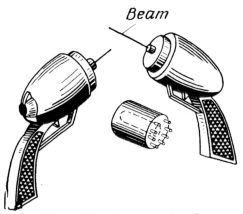

Fig. 82. Laser pistol

example, it is reasonable to employ pulsed-working ruby lasers (Fig. 82).

In this case the excitation source is made as an easily replaceable cartridge. The ruby rod is arranged along the axis of the cartridge together with battery-powered chemical excitation sources.

A laser rifle for the US army has been developed by the Maser Optics Co. The weight of this rifle is 11.3 kg. It is powered from a storage battery which ensures 10 000 flashes. The rate of firing is one flash every 10 seconds.

Such weapon as a laser pistol or a laser rifle can injure the eyes of man. Powerful radiation concentrated on the retina by the crystalline lens causes its damage. The laser radiation pulses being very short, the organism has no time to defend them.

Published reports, however, hold a laser rifle or pistol to be not sufficiently effective destruction means, since a laser beam can affect visual organs only when the "victim" is looking in the direction of the enemy. Even slight fog or smoke may substantially diminish the destructive action of such weapon.

For protecting the eyes of man against the action

of the laser light beam investigations and tests are reported to be carried out in the United States with photochromic solutions which, when exposed to light, are capable of instantaneously changing the extent to which they are coloured and become non-transparent. A photochromic solution consists of a solvent, a photochromic dye, and an enzyme (ferment) that controls the speed of the reaction, i.e. of colouring. Such solutions are a thousand times more responsive to intensive visible and ultra-violet radiation than the human eye. The colouring of a photochromic solution to a high optical density takes 10 μs. After the intensity of the flash has dropped down to the tolerable level, the solution in a few milliseconds becomes transparent again.

For protecting the eyes against a light flash, such a solution is used to fill the space between the two transparent component parts of a motor-car windscreen, of a protective mask facepiece or eyepieces.

LASERS IN MEDICINE AND BIOLOGY

Remarkable properties of lasers attracted the attention of surgeons. A laser beam proved to be fit for performing such operations as are usually done with a scalpel. In this case a laser beam from a generator is transmitted along a flexible light guide made as a bunch of glass or plastic fibres. The light guide terminates in a lens and is provided with a handle for the surgeon to manipulate it. The lens focuses the light beam into a small spot with a diameter of several angströms. With such a scalpel, it is possible not only to incise tissues being sure of complete sterility, but also to dissect individual cells.

The laser beam will be employed as a cauterizing instrument for treating skin neoplasms and injuries.

In this case a laser is advantageous over conventional cauterizing means, since it is an absolutely sterile instrument and will not introduce any secondary infection.

But, probably, lasers will be most valuable instruments in eye surgery. The fact is that a laser beam of a definite intensity can pass through the transparent tissues of the eye without injuring them, so that operations on the eyeground can be performed without causing the patient to suffer poignant pain. The flash lasting but for a short period of time, overheating of the eye or its injury in case of an involuntary contraction of the eye muscles are precluded.

Numerous successful operations on the eye have already been performed with the help of lasers. We shall cite only two examples. Using a laser as a surgical instrument, it was possible to remove a tumour from the ophthalmic artery; with a laser functioning as a photocoagulator, detached retina was successfully "welded" to the eyeground.

This method of medical treatment has received most serious attention in the Soviet Union, and special lasers for medical purposes have been created here. Most of the operations performed at the V. P. Filatov Institute of Ophthalmology in Odessa with the use of lasers were successful, so that quite a number of patients could see again. The new method is the subject of comprehensive research at other Soviet clinical institutes as well.

Figure 83 is a photograph of one of laser ophthalmocoagulators developed in the Soviet Union and designed for carrying out complicated operations on the eye. The radiation source in this apparatus is a ruby rod, 6.5 mm in diameter and 65 mm in length. The laser gives 4 flashes a minute, each lasting for 1 or 5 ms. The pulse radiation energy ranges from hun-

Fig. 83. Soviet laser ophthalmocoagulator Model OK-1

dredth fractions of a joule to 1 J. The minimum diameter of the light spot is about 100 μ. The apparatus is powered from commercial mains (220 V). Its power consumption is 300 W.

This is, certainly, not the complete list of possible medical uses of lasers. Lasers will find application in cancer research and treatment. Some specialists consider that lasers will usher in a new era in cancer surgery. The first experiments in the laser treatment of malignant tumours carried out on animals and man gave, though not yet conclusive, but encouraging results. One of such experiments was conducted with a ruby laser ($\lambda = 6943$ Å) and with a gas laser ($\lambda = 6328$ Å).

The experiment was run on nine animals (Syrian golden hamsters) to which amelanotic melanoma was inculcated from man. The animals were subjected to laser irradiation, and after such treatment the tumours disappeared in all the nine of them. One month later no traces of tumour could be detected even by microscopic investigations.

The radiation energy of a ruby laser in this experiment was within a range of 60 to 380 J, and the pulse power was 100 MW.

Tumours of other type, e. g. transplanted fibrosarcoma, are less sensitive to laser radiation, and in some cases they could not be destroyed.

Similar experiments were carried out with human patients. One such patient suffered from a malignant tumour (melanoma) with metastases developed into the skin and subcutaneous tissues; the metastases reached 1 cm in diameter. All these tumours were irradiated with a series of pulses, their total energy coming to 360 J. Twenty days after the irradiation the tumours disappeared. The effect was more pronounced when the tumour was exposed to focused radiation, with a sufficient density of the energy incident on the tumour surface area. Thus, in the experiment under discussion the energy was focused into a 2 mm-diameter spot and the energy density was 1500 J/cm^2.

In certain cases the effect of the treatment can be enhanced by staining the tumour surface and thus increasing the energy absorption coefficient.

The problems of laser applications for treating malignant tumours are studied by the National Institute of Cancer in the United States as well. One of the models of a medical laser developed by this Institute has an output energy of 800 J, with the pulse repetition rate of 4 p.p.s. and the pulse duration varying from 2 to 4 ms. In appearance, this laser resembles a dentist's drill. It is accommodated in a shuttle suspended from the ceiling. A set of lenses, prisms and mirrors housed in a connection sleeve serve for transmitting the energy to a small instrument that the surgeon has to manipulate directly above the area to be operated. With the aid of the lenses the surgeon can vary the radiation energy within 100 to 800 J. Energy losses during its transmission along the sleeve do not exceed 8 per cent.

The laser consists of four heads, each of them being a Pyrex cylinder with a rod from neodymium-activated glass. The rod length is 91 cm and its diameter is 1.9 cm. The cylinder is filled with water for cooling. The laser rods are pumped by a 5 kW flash tube.

Certain success was reported in laser therapy of superficial malignant tumours. It was also reported that such a laser could be used as an auxiliary means in cancer surgery when removing neoplasms in such organs as liver and lungs, where the application of a conventional scalpel involves a considerable risk, as well as for destroying the tissues surrounding the main tumour, since it is not always possible to remove all the peripheral neoplasms without running the risk that the patient's life might be lost.

Experiments with animals showed that implanted tumours could be rapidly destroyed without affecting the neighbouring sound tissues, provided the laser radiation is appropriately focused. Some researchers consider that the laser radiation effects on normal and malignant tissues are different, especially when the malignant tissue is pigmented.

Extensive investigations are made into the biological effect of lasers on living organisms, particularly on individual cells and on the central nervous system. In experiments with mice, scientists were able to cause severe lesions of different parts of the brain by means of focused laser radiation. Depending on the accuracy with which the radiation was focused, different parts of the brain could be destroyed, starting with the cortex and down to the deep-lying strata— the white substance of the spinal bulb and its other parts. A remarkable fact in this operation was that the cranial bones and the pachymeninx (dura mater) remained intact. During the experiment the animals were irradiated by a ruby laser beam, with the pulse energy not exceeding 40 J. The distance from the end face of the laser to the animal's head was 2 m. Most of the animals perished during the experiment. The lesion was caused, evidently, by a sharp temperature increase in the point of the beam focusing.

Research in this direction is continued.

CHAPTER 5

Lasers and Science

TESTING EINSTEIN'S THEORY OF RELATIVITY

One of the postulates on which the special theory of relativity developed by Albert Einstein in 1905 is based reads that the velocity of light in vacuum is a constant value, the same in all inertial reference systems and equal to

$$c = 299\,792.5 \text{ km/s}$$

In other words, the velocity of light is dependent neither on the movement of the source, nor on the movement of the observer (receiver).

This postulate was inferred from the negative result of the celebrated experiment carried out in 1887 by A. Michelson and E. W. Morley with a view to demonstrating the existence of an "aether drift". They wished to find out whether the velocity of light varies, and if it does, then how, in case the observer moves towards the source of light or away from it.

It is not at all easy to detect a difference in the light propagation velocity, taking into account how tremendous this velocity is and how small the stations required for the experiment are. A particular accuracy is needed in this case. Nevertheless, A. Michelson and

E. W. Morley managed to overcome all the difficulties by basing their experiment on the wave properties of light.

The apparatus used in the experiment was an interferometer with a multiple reflection of the light beam. In this experiment the time difference in the beam travel in different directions was to be equal to $0.4 \cdot 10^{-15}$ s (this being provided by turning the interferometer frame freely floating on the surface of mercury poured into a special vessel through 180°). Since the period of light oscillations for visible rays is 10^{-15} s, the time difference will be 0.4 of the period. Proceeding from the interference of the oscillations of the first and second rays, the phase difference in these oscillations could be determined with an accuracy of up to 0.01 of the period.

Thus the error in observing the interference pattern in the Michelson—Morley experiment was only 2.5 per cent. And still, in spite of a comparatively high accuracy, the results of the experiment were negative. No time difference in the propagation of the rays could be detected with the interferometer in either of its two positions. In other words, no "aether drift" was shown to exist. The velocity of light remained constant, irrespective of whether it travelled in a direction towards the moving object, or away from it.

This negative result of the Michelson—Morley experiment created a difficult situation for physicists as it was in contradiction with the theory of the aether at rest.

As the way out of this situation, Lorentz and Fitzgerald put forward a hypothesis, according to which moving bodies contract in the direction of their motion. This effect, known as the Lorentz—Fitzgerald contraction, was also used by A. Einstein in the creation of his special theory of relativity.

Gas lasers allowed the Michelson—Morley experiment to be repeated with an accuracy about one thousand times exceeding that attainable before. The idea was as follows. The working frequency of a gas laser must be determined by the direction of propagation of light in the resonant cavity relative to the aether drift. A change in the light propagation velocity in the resonant cavity (or a change in the length of the latter) would be detected as a variation of the laser frequency.

One possible scheme of this experiment with the use of lasers is shown in Fig. 84. In position a laser A radiates in the direction of the supposed aether drift, and another laser B radiates in a direction perpendicular to this drift. The signals of these two lasers are mixed by a photodetector, and a certain frequency difference signal is obtained at its output. In position b one of the lasers (in this case laser A) radiates in a direction opposite to the supposed aether drift. The other laser (laser B), as in position a, radiates in the direction perpendicular to the aether drift. At the output of the photodetector we shall again have a certain signal equal to the frequency difference of the radiations of lasers A and B. Should an aether influencing the light propagation velocity exist, this would cause a shift of the frequencies of one of the lasers, A, and this shift would be different from the frequency difference signal at the detector output in position a. But no difference was observed in the experiment. It was established that with a difference in the orientation of the laser with respect to the Earth and to the supposed aether drift its working frequency remains unchanged. This experiment demonstrated the absence of variations in the velocity of light with an accuracy of up to 0.03 mm/s.

Thus the validity of Einstein's theory of relativity was tested once again by means of lasers.

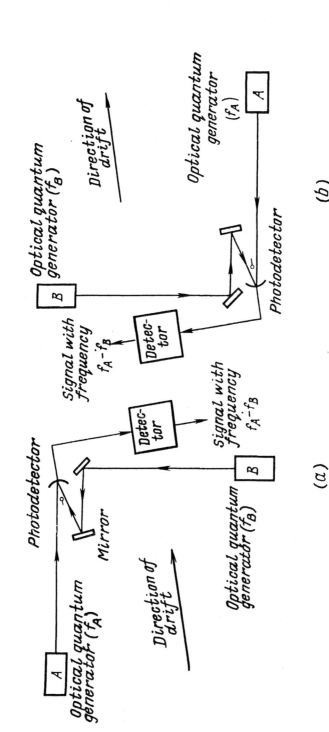

Fig. 84. Diagram of an experimental system for testing the theory of relativity

MEASURING THE DRIFT OF CONTINENTS BY MEANS OF LASERS

There exists a hypothesis which maintains that the position of the continents on the globe is not always the same, but varies with time. This phenomenon is termed drift of continents. According to the latest assumptions, the continents drift at an average speed of 2 to 5 cm a year. But how can this unusual hypothesis be verified? Do the continents remain immovable or really travel along the sphere of the Earth? And if they do, then how fast?

Yet, there is a possibility of subjecting this hypothesis to an experimental test with the use of lasers. A helium-neon CW laser will, probably, be most fit for conducting such an experiment, since the stability and coherence of a gas laser radiation are quite adequate for meeting the experimental requirements. It will be recalled here that the length of one of the laser radiation waves is 1.135 μ. The frequency of the electromagnetic field oscillations is approximately $3 \cdot 10^{14}$ Hz.

Figure 85 shows the scheme of an experiment during which the Doppler effect manifestation may be expected.

Let the laser be found at point *1*, and its beam be directed onto a mirror reflector located at point *2*, travelling with a velocity v in relation to the laser (point *1*). If the reflected light is mixed on a photomultiplier with a portion of the light initially produced by the laser, components arising from the Doppler shift must appear at the photomultiplier output. The frequency shift Δf is given by the equation

$$\Delta f = N \frac{v}{\lambda}$$

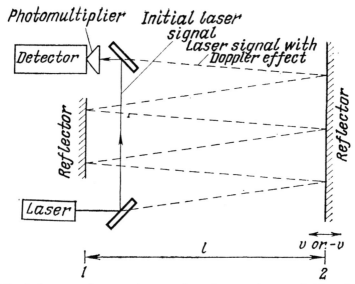

Fig. 85. Scheme of experiment for determining drift of continents

where N is the number of the beam passages between points *1* and *2*; v is the velocity in angströms per second; and λ is the wavelength in angströms.

The average speed of 5 cm/year corresponds to approximately 20 Å/s, i.e. the displacement per second is approximately equal to 20 atomic diameters (one angström corresponding to the diameter of a hydrogen atom).

The horizontal continental drift is considered to be conditioned by convection water currents in the Earth, entraining the continents. Those structural discontinuities into which the convection currents vanish or from which they come close to the Earth's surface may be the cause and locality of the continental drifts. There are such places where large masses of land border these discontinuities from both sides and the distance between them is not very great (about 30 kilometres). The Straits of Bab-el-Mandeb and Gibraltar can be mentioned as examples.

These regions are, perhaps, most suitable for carrying out the above-described experiment and verifying the continental migration hypothesis. In such a case points *1* and *2* should be located on the opposite sides of the discontinuity. The Doppler shift per passage in both ways corresponds to 1/250 Hz, and this value cannot be detected against the background of casual short-time shifts of the laser frequency. The Doppler shift can be extended by a multiple reflection of the laser beam in reflectors located on the opposite sides of the discontinuity. Assuming that no losses take place, the distance of 15 km the light can traverse 500 times in both directions, till the coherence disappears. This will give a 500-fold increase of the Doppler shift, i.e. $500 \cdot 1/250 = 2$ Hz.

With a continuous lasing power of 50 MW, the reflection coefficient on the mirrors of 0.9 and losses in the medium of 3.0 db, a photodetector having a conventional sensitivity will be able to detect a 20-fold passage of the beam in both ways. The Doppler shift then will be 0.1 Hz.

If more powerful lasers are used, the Doppler shift may be increased, since a greater number of beam passages between the reflectors will become possible before the signal level has become minimum detectable. Besides, an increase in the coherence of the source can still further improve the frequency stability. With the continuous lasing power of the order of 10 W, the Doppler shift could be slightly increased (to 0.11 Hz altogether), and the short-time stability of the laser frequency could be diminished to 0.1 Hz. The Doppler shift magnitude being determined, the continental drift velocity can be found as

$$v = \Delta f \lambda / N$$

The realisation of such an experiment is naturally associated with certain difficulties. These are as follows.

If the continental drift velocity is actually 5 cm a year on an average, then the movement is, most likely, non-uniform. However, an intermittent movement is easier to measure, for the speed at the moment of shocks will greatly exceed that of a uniform motion. The velocity and magnitude of drift are possibly dependent on temperature, pressure and tides which vary with seasons and during the day.

The Earth is subject to shocks and jerks, it has a kind of pulse of its own. All these factors will tell on the experiment, if run as suggested above.

Different variations in the phase characteristic of the signal will be caused by casual fluctuations in the refractive index of air. Therefore the distance l cannot be very great. The determination of the magnitude of this effect requires experiments on the propagation of laser light over great distances.

And finally, when conducting such an experiment, definite requirements must be met as to the stability of the foundations on which the instruments will be located at points 1 and 2, so as to preclude local vibrations of the ground.

All these difficulties can probably be surmounted. Experiments based on the principle of the multiple reflection of the beam have already been carried out (such as the Michelson—Morley experiment discussed above). Under the conditions of continuous oscillations, correlation methods can be used for data processing.

The above-described experiment on measuring the continental migration velocity may prove to be a valuable tool for geodetic explorations of the Earth's surface. The same idea can be used for measuring the rate of glacier movement.

LASERS FOR GEODETIC STUDIES AND ATMOSPHERIC SOUNDING

Lasers will find application in diverse geodetic instruments. The use of lasers on seismographs is worthy of attention. A seismograph whose block diagram is shown in Fig. 86 consists of two gas lasers *4*. One of the mirrors of the resonant cavity *3* of each laser is connected with an oscillatory mass *2* suspended from spring *1*. The other two mirrors *5* are stationary. With the pendulum swinging, the length of one resonant cavity increases and that of the other diminishes. The result is a corresponding alteration in the working frequencies of the lasers (f_1, f_2). With the aid of optical mixer *6* the frequency difference (f_1-f_2) is produced, whose variation corresponds to the amplitude

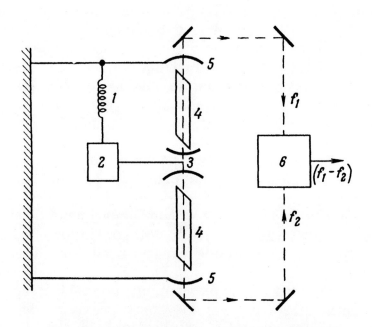

Fig. 86. Diagram of a laser seismograph
1 — spring; *2* — oscillatory mass; *3* — movable mirrors; *4* — gas lasers; *5* — stationary mirrors; *6* — mixer

of the pendulum oscillations. The sensitivity of such a seismograph is at least by one order of magnitude better than that of other types of seismographs.

The laser can be employed for atmospheric sounding to determine the height and density of the metastable states of the upper atmospheric layers, caused by such phenomena as aurora, airglow, solar eruptions, meteor trails.

The whole system employing a laser for the above purposes may be called an optical radar. It consists of an emitter of an intensive coherent light pulse and a receiver which measures the amplitude and delay time of the reflected optical signal. A pulsed gas laser is most fit as a transmitter, and a photomultiplier is preferable as a receiver. Since each atmospheric component is best sounded by pulses of a definite wavelength, a laser is evidently needed whose radiation frequency can be retuned, otherwise, the system should include several lasers, each of them tuned to a definite component of the upper atmospheric layers.

For determining the concentrations of metastable nitrogen, for instance during the aurora period, as well as for ascertaining its spatial distribution in the atmosphere, a high-power pulsed working nitrogen laser could be employed.

By adding molecules of sodium into the active medium of a pulsed working gas laser, a sodium laser can be created which will emit waves belonging to the yellow region of the optical spectrum. Such sodium laser would allow examination of sodium density distribution in the atmosphere in a vertical plane during the whole night. The laser makes possible the detection of turbulent air flows in the atmosphere.

Calculations show that the development of such a system is possible and that it will be an extremely valuable tool for atmospheric sounding.

MEASURING OF SPEEDS

By using a laser, speeds can be measured with a high accuracy. The relative speed of an object in such a case is found from the formula

$$v = \frac{\Delta f c}{2f}$$

where v is the speed to be measured; c is the velocity of light; f is the laser frequency; and Δf is the frequency shift due to the Doppler effect.

Gas lasers, ensuring a high degree of coherence and monochromaticity, are preferable for this purpose. The frequency of lasers operating within the visible or near infra-red regions of the spectrum is known. Let it be 10^{14} Hz. If the speed of an object is several centimetres per second, the frequency shift will be about 10 kHz. Thus, both very small and high speeds can be measured with a high accuracy.

LASER SPACE COMMUNICATIONS

In spite of certain difficulties, even to-day the use of lasers for space communications can be said to be extremely promising.

Let us consider some of the problems associated with the possibility of using lasers for space communications. The works of B. Bowers are devoted to these problems.

Some recapitulation concerning the structure of light will, perhaps, be appropriate here. Any light source radiates light not continually, but as discrete "clusters" of energy, called light quanta, photons. The energy of a photon depends on the radiation wavelength and is the greater, the shorter the wavelength. Therefore it can be said that a photon corresponding to the ultra-violet region of the spectrum possesses a higher energy than a photon in the infra-red region. Ultra-violet rays are characterised by the shortest wavelength of the optical spectrum, while infra-red rays, on the contrary, are characterised by the longest wavelength.

The quantity of photons emitted by a system per minute is a very important magnitude for characterising the quality of communication. In space communication system literally every photon must be taken into account. The quantity of photons eventually determines the information transmission rate.

In order to ensure maximum range of optical communication, it is obviously necessary to bring the energy density in the light beam to the highest possible degree. An increase in the density of energy in the beam requires an increase in the dimensions of the transmitting aerial. From the Fraunhofer diffraction theory it is known that the angular width of a light beam must be proportional to λ/d where λ is the radiation wavelength and d is the diameter of the transmitting aerial. For this reason even microwave aerials are bulky. When we are to deal with a communication system designed to operate in the optical range in the outer space where such factors as weight and dimensions of the equipment are of primary importance, the advantages of lasers over conventional radioengineering means are indisputable.

In any communication system so-called background noises are present. Usually these noises are from both external and internal sources (electron valves, resistors, power supply sources). The quality of the received signal depends on the level of these noises and is characterised by the signal-to-noise ratio. The main potential source of internal noises is the transmitter which must have a very low level of intrinsic noises. One of the most important characteristics of the laser is just a low level of its intrinsic noises, as compared to external sources. The external source of noises is the solar and lunar chromosphere. As is known, radiators are characterised by the temperature related to that of a black body. For the Sun this temperature is estimated at approximately 7000°K. The influence of the Moon is less significant, but also should be taken into account. The effective temperature related to that of a black body for the Moon within its areas illuminated by the Sun is estimated at 407°K. The average value of the background received on the Earth is only 273°K.

If the receiver is screened against sunlight, the background noise caused by other sources is insignificant. Therefore optical communication is better during the night than in the daytime. This is true for the infrared region as well.

The intrinsic radiation spectrum of the Moon is shifted towards longer waves. The minimum spectral emittance of the Moon (minimum energy density per unit wavelength of radiation) corresponds to the waves having a length of about 10μ.

When choosing the frequency for laser communications, certain limitations should be taken into consideration. For instance, the quantity of photons emitted by the source at a pre-set power level is the greater, the greater the wavelength. Besides, one should try to exclude background radiation, and this is not at all easy to do in the optical range. Another limitation is associated with the absorption of radiation by the terrestrial atmosphere, so that "optical windows", that is, such wavelength regions for which the atmosphere is almost transparent, should be sought for. All these requirements taken together make the wavelength region of about $10\ \mu$ most fit for optical communication systems.

As we know, powerful CW lasers operating in this very range have already been developed.

Perhaps, the most difficult problem in the use of optical communication systems in the outer space is the creation of superstable laser platforms. On the one hand, the width of the beam radiated by the emitter must be rather small and, on the other hand, maximum stability of the platform mounting the radiation source is required, so that angular fluctuations of the laser beam about the line connecting the transmitter with the receiving station on the Earth should be comparable with the angular width of the beam. The same contradictory requirements are to be met by the radiation receiver. On the one hand, the receiver must be able to receive maximum power that reaches it when the receiver fully covers the entire cross-section of the radiated beam; on the other hand, the dimensions of the receiver aerial must be as small as possible, so as to minimise the reception of background radiation. Consequently, the platform of the transmitter must be stabilised in the best possible way.

The width of laser beams is on the order of 10^{-4} radian (which corresponds to approximately 200"). As to the stability of presently operating laser systems, it lies so far within several tenths of a degree and may reach even 5°, depending on the magnitude of the forces acting on the platform (solar radiation pressure, cosmic magnetic fields, inertial forces, etc.). Because of this, the region covered by the radiation of lasers in the outer space in the present-day state of engineering turns out to be at least 1000 times greater than the cross-sectional area of their beam on the Earth's surface on account of instability of laser platforms (it should be remembered that it is the axial angle which is meant here, but not the solid one).

A compromise, though a poor one, would be to employ several lasers simultaneously transmitting the same information.

Due to the fact that the axial rotation velocity of the Moon is less than that of the Earth and that there is no atmosphere on the Moon, the transmission of information from the Moon to Mars appears to be more advantageous than such a transmission from the Earth, though attenuation of radiation in the Martian atmosphere should be taken into account.

Using such a stable platform as the Moon in relation to the Earth, it will be most simple to effect a one-way Moon-to-Earth communication, and therefore the first experiment of a laser interplanetary communication will, probably, be conducted between the Moon and the Earth.

It is most likely that a research station will be erected on the Moon. It is very probable that a laser transmitter will be mounted on this station, with its powerful beam directed towards the Earth. The bright laser beam can then be seen with a naked eye even against the sunlit side of the Moon. When the Moon faces our Earth with its dark side, the laser beam will flash up against its background as a dark-red star (its colour, naturally, depending on the selected frequency of the laser), surrounded by a golden crescent at new moons.

Very simple messages, of interest to the compatriots of the astronauts and to the entire population of the Earth, could be transmitted by periodically interrupting the laser beam by means of a screen (modulating it). For transmitting more complicated messages and larger volumes of information, the laser beam must be modulated with the use of a different, more perfect method. Let us determine the optimal width of a laser beam, which would ensure both the required density of the beam energy and reliable capturing of an object to which information is to be transmitted.

As is known, a light beam is scattered on account of the inhomogeneity of the atmosphere. Besides, diffraction phenomena add to the light beam scattering and bring the scattering angle to about 5 angular seconds. All this will result in an appreciable deterioration of laser characteristics.

What is the relationship between the width of a laser beam and the power of a laser?

This relationship is expressed by a complicated formula; we shall not consider it here, but merely state that the power at the point of reception is determined by the relation of the area of the receiving aerial to the area a laser beam would cover in a plane perpendicular to the direction of its propagation and passing through this point. Evidently, this area will increase in proportion to the square of the radius of this plane (assuming it to be a circumference). The width of the beam (its angle) is defined as the relation of this radius to the distance between the laser and the receiver. This, naturally, is only a rough approximation. Many other factors such as the signal-to-noise ratio, signal bandwidth, spectral density of noises, conditions of the atmosphere or medium traversed by the beam, etc. also enter into the formula.

In order to practically estimate the possibility of establishing communication between the Earth and the Moon it is necessary to know the signal-to-noise ratio. Calculations show that for a laser working at the wave of 10 μ, with a laser beam width of 20", the surface area of the receiving aerial of the reflector of 20 m^2 and the laser output power of 0.1 W, the number of photons received by the aerial will be $2.25 \cdot 10^{11}$ photon/s. The number of photons constituting the noise background, according to calculations, is $8 \cdot 10^{10}$

photon/s. Consequently, for the adopted values the signal-to-noise ratio will be

$$\frac{S}{N} = \frac{2.25 \cdot 10^{11}}{8 \cdot 10^{10}} \approx 3$$

Insofar as we have taken into account not all the interfering factors, the actual value of this ratio will probably be smaller. However, the output power of the now existing lasers, especially of ruby lasers, is considerably greater than 0.1 W adopted in our calculations, and therefore the signal-to-noise ratio may practically be better. Yet, it should be taken into consideration that an increase in the laser power will entail broadening of the laser output spectrum.

An appropriate choice of the type of a detector is very important for the construction of optical space communication systems. Photomultipliers, laser receivers and quantum counters may be used for this purpose.

Photomultipliers have a very broad radiation passband, so that in detectors employed in space communicaton systems narrow band-pass optical filters must be used which would pass only one narrow and definite frequency band. This band must fit the atmospheric transmission windows (of the terrestrial atmosphere for the case of a Moon-to-Earth communication system) and the laser transmitter frequency.

At the same time photomultipliers have a merit of possessing a large capture angle (so that the direction of radiation may mismatch the photomultiplier axis to a considerable degree). Proper cooling of the photomultiplier improves its performance characteristics and brings down its noise level. But the level of external noises (background and others) still remains high. The elimination of external noises is the main problem with photomultipliers, and its solution may be found in modulating the light beam.

A laser receiver (i.e. a laser functioning as an amplifier) will be difficult to employ because of small capture angles. If we recall the laser design, it will be easy to understand that only those photons which have entered the laser in a direction strictly parallel to the axis of its rod will cause generation and be amplified; photons which fall at an angle to the rod axis do not cause generation, leave the rod at an angle and are scattered. Collecting optical systems which focus the bunch of received photons onto the ruby improve the operation of a laser receiver. However, this type of a detector suffers from an essential disadvantage on account of considerable fluctuations in the number of the signal photons received by the detector per second. Besides, a laser amplifier has a relatively broad photon capture band, i.e. is triggered by photons having different energy and frequency.

A quantum counter is most effective for detecting the signal frequency within a broad band of frequencies. The operation of this counter is based on the use of strictly definite energy levels. For revealing a transition between these levels, the frequencies of the absorbed and radiated light must be the same. This principle is realised in the laser. The difference between a quantum counter and a laser resides in that the former responds to strictly definite frequencies.

Another possible method of detecting signals is the so-called method of photon statistics. If the power of a coherent radiation is commensurable with that of a fluctuation interference, then the detector during a certain small period of time will be unable to discriminate between the signal and interference. The detector will record a signal only after a change has taken place from the random fluctuation towards a definite value. With this method, the information transmission rate will evidently be rather low.

Let us calculate, how many photons a detector will require for discriminating signal photons against the background of noises. Let us assume the signal recognition criterion to be the level at which the signal is at least equal to background fluctuations. Then in case of a purely random arrival of photons to the detector surface the number of photons is $8 \cdot 10^{10}$ photon/s. Assuming that the filter passes radiation only within the laser emission band (optimal case), the number of photons reaching the detector will be $3 \cdot 10^5$ photon/s. By multiplying this value by the energy of a single photon with a wavelength of 10 µ, we shall obtain the power of a minimum signal at the detector to be on the order of $6 \cdot 10^{-15}$ W. According to experimental data this minimum power is approximately 10^{-12} W.

In order to realise these communication system projects much progress is to be made in the field of optics. Nevertheless, the potentialities of optical communication systems are extremely high, and this is confirmed by the reception on the Earth of the laser beam sent to the Moon and reflected from its surface.

Another important aspect of laser space communications is the provision of communication between space ships and some planet or the Moon. For the solution of this problem, the stabilization of the laser platform (space ship) must be improved by at least hundreds of times. At present it is very difficult to solve this problem since the forces acting upon an artificial satellite, such as pressure of sunlight, force of gravity, collision with various particles, etc., are not easily determined. Progress in this field will depend on the solution of the main problem of aiming at and tracking of space objects.

Calculations of theoretical possibilities of a commu-

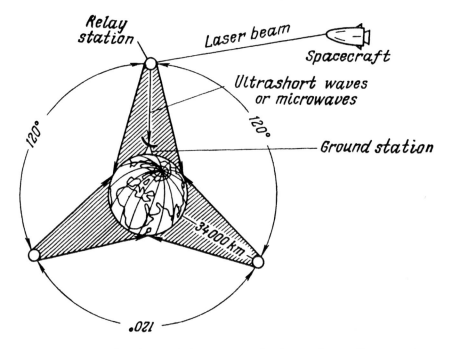

Fig. 87. Communication with spacecraft through a relay station

nicaton system in the outer space with the help of lasers allow one to come to the conclusion that such systems will be created.

The realisation of one of the possible projects of space communication between the Earth and Mars with a relay station on the Moon contemplates taking advantage of the absence of lunar atmosphere and a higher stability of such a "platform" as is the Moon in relation to the Earth. This scheme may also be used for establishing communication with spacecraft in flight.

In those cases when a planet has no suitable natural satellite on which a relay station could be mounted, it is reasonable to mount relay stations on artificial satellites especially for communications with spacecraft. A scheme of such a communication system is

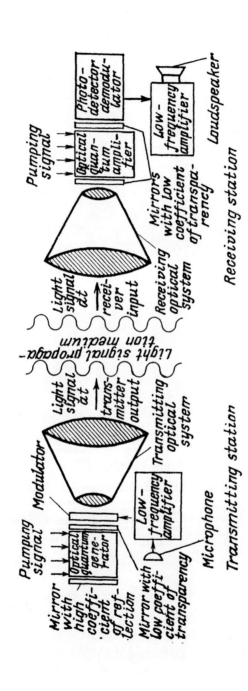

Fig. 88. Communication system with an optical quantum generator (transmitter) and an optical quantum amplifier (receiver)

shown in Fig. 87. The artificial satellite will be in orbit outside the atmosphere. If the atmosphere of the planet is sufficiently dense, the transmission within the planet-satellite region may be better carried out using UHF or SHF bands.

For a reliable communication between Earth stations and spacecraft, it is also expedient to employ stationary satellites serving within a global communication system which will evidently be created in the nearest future. With this system, one of the satellites will always be within the zone of assured communication with one of the spacecraft. Two other satellites will ensure the relaying of the messages from the spacecraft to any point of the globe.

When realising the space communication projects, such a factor as energy requirements for the transmission of information are to be taken into account. Calculations show a laser system to be most economically expedient in this respect, since it requires only 10^{-16} W/s for the transmission of one binary digit of information, whereas a system operating in the radio frequency band requires 10^{-7} W/s for the transmission of the same amount of information. A laser system thus requires one thousand million times less energy than a radio frequency one.

Figure 88 shows a version of a communication system in which the transmitter is a laser functioning as a generator and the receiver is an optical quantum amplifier. Light radiation is modulated by audio frequency signals produced by a low-frequency amplifier. The transmitter is provided with an optical system that shapes a beam and directs it towards the receiver. The optical system of the receiver is in alignment with that of the transmitter. The light signal comes to the optical quantum amplifier. On arrival of the light beam the optical quantum amplifier am-

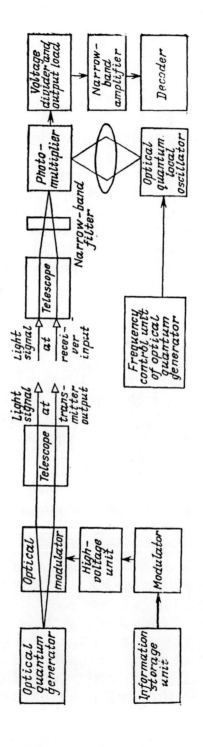

Fig. 89. Diagram of a version of a laser communication system

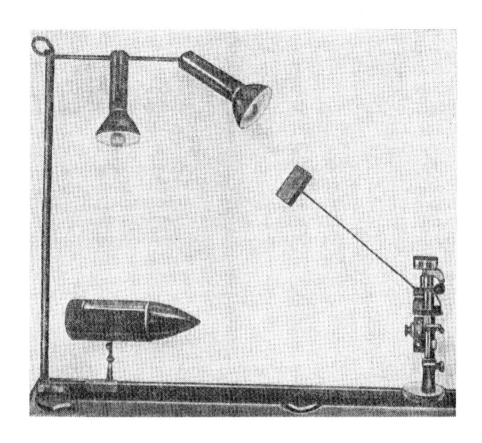

Fig. 90. Mock-up of a laser space communication system

plifies the signal due to induced radiation, and the amplified signal comes to a photodetector and a demodulator. From the demodulator the signal is sent to the low-frequency amplifier. A specific feature of this system is that it is almost insensitive to interferences, since only those signals are amplified which come to the input of the optical quantum amplifier strictly parallel to its optic axis.

Figure 89 shows a version of a communication system also based on the use of lasers, but differing from the preceding one in that for mixing the bunches of light from the transmitting laser and from the local oscillator laser it envisages the use of a photocell functioning as a demodulator. When mixing the light bunches, this demodulator separates the radio-frequency signal which carries the transmitted information. The operation of other units of this system is clear from the Figure.

Figure 90 shows a mock-up of a system which demonstrates the possibility of effecting space communications with the use of laser radiation and employing solar energy. A semiconductor laser placed into a cryostat (to the right) is powered from solar batteries. Two lamps at the top simulate the Sun. Seen to the left is a mock-up of a space vehicle bearing a laser radiation indicator. The laser emits 400 pulses per second of 1 μs each. The indicator converts light signals into sound signals.

The possible versions of various space communication systems with the use of lasers are certainly not limited to these examples.

CHAPTER 6

The Prospects of Lasers

PIPELINE OUT OF A LASER BEAM

There is still another interesting application that lasers may find in the future: to convey gaseous substances, say, oxygen or hydrogen, over interplanetary distances, with the laser beam put to use as a pipeline. Such a perspective is discussed by G. Pokrovsky in his book "Superpowerful Light" (Znaniye Publishers, 1964).

If we take a laser shaped as a hollow cylinder, its outgoing beam will be tubular. The light flux in this tubular beam is so distributed that the density of the energy in the mid-portion is lower than in the peripheral layer (Fig. 91).

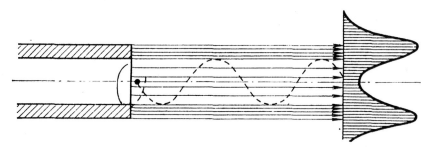

Fig. 91. Diagram of a pipeline out of a laser beam

Now, if we introduce a certain amount of gas into the mid-portion of the light flux, this gas will be entrained by the light flux inside the light tube in the direction of propagation of the light. The molecules or atoms of the gas will be unable to escape through the peripheral zone since it has a higher concentration of the energy. Thus, a kind of a gas pipeline is created by the light rays which retain the gas and make it travel with the light wind. Unlike a conventional pipe, this light pipe will offer no hindrance to the passage of gas particles, there being no friction against its walls. On the contrary, the walls of the light pipe will add to the effect of the light wind.

However, on account of imperfect finishing of the optical surfaces of the radiating system and because of the presence of diffraction phenomena, the light pipe is effective only over a certain distance.

Suppose it is necessary to convey a gaseous substance along a light pipeline from the Earth to the Moon. What will be the diameter of the transmitting device? According to calculations, with the standards so far reached in the manufacture of optical devices, the diameter must be 89 metres. At present such an emitter can hardly be constructed; but with further advances in the technology of optics this figure may be reduced to 25 metres.

For intercepting the substance transferred along a light pipeline, a corresponding receiver will be needed with a diameter twice that of the emitter, that is, 198 metres with our present-day techniques or 50 metres with more advanced ones.

The axis of the light flux must naturally be strictly oriented towards the receiver. For effecting the transfer of a gaseous substance along a light pipeline "routed" between the Earth and the Moon, the axis of the light flux must follow the motion of the Moon, and this can be achieved through the use of a special turning servo-mechanism operating with a very high accuracy due to the provision of a feedback. The feedback will give information about the location of the light spot on the Moon in relation to the receiver and allow the transmission of appropriate correction signals to the Earth. The arrangements of such kind are quite feasible at present.

When realising the projects of transmitting substances along light pipelines, it should be taken into account that hydrogen, oxygen and other gases rather poorly absorb and scatter light under normal pressure and temperature; therefore they are poorly entrained by the light flux. To improve the situation and raise the gas transfer velocity, we may first heat and ionise the gas, bringing it to the plasma state. If the temperature is raised to 5 or 6 thousand degrees, the gas will intensively glow and absorb light. In such a state it will be easily made to move under the effect of the light flux pressure. Though very powerful radiators would be required for delivering gaseous substances over tremendous distances, the procedure itself appears rather realistic.

In his book G. Pokrovsky points out that power requirements for satisfying the needs in oxygen of one man on the Moon will be at least several dozen thousands of kilowatts. This means that radiating devices controlled with an extremely high accuracy must be as big as dozens of metres.

All this may seem phantastic. But man is steadily pursuing the task of conquering the nearest planets, and in so doing, it will take him, perhaps, only a few decades to make the projects of transferring gases along light pipelines a reality.

LASERS AND COMMUNICATIONS WITH EXTRATERRESTRIAL CIVILIZATIONS

Are we alone in the Universe? Is man the sole intelligent being to enjoy the privilege of contemplating the splendour of the Nature, or other intelligent beings similar to us, men, exist somewhere in the immense Universe? And if they do exist, how can we get in contact with them and reach mutual understanding?

Ages ago man regarded himself to be in the very centre of the Universe and acting alone on its boundless scene.

But then came the discovery that a great many other worlds surround both our Earth and Solar System. This, naturally, suggested that life could possibly exist somewhere else as well.

Why not on the Moon, our nearest celestial neighbour? This idea, however, was to be abandoned, since the Moon turned out to be deprived of the atmosphere, and man directed his attention to Mars and Venus. The thought that these planets might have intelligent inhabitants gave birth to a wealth of legends told and fiction stories written. Yet, recent explorations of these planets leave us less and less hope that intelligent beings may live on either of them. Scientists attribute seasonal changes in the colour of the Mars' surface to the presence of certain forms of vegetable life there; but the final say will be, before very long, when man has made his first steps on the mysterious planet.

The Universe has no boundaries and the Solar System is just a speck in it. Indeed, in our Galaxy, a family of about 100 000 million stars, some stars have their planets, as our Sun does. We cannot see these planets even through most powerful telescopes, but investigations into the behaviour of these stars confirm the existence of their own planets.

Suppose now that out of a thousand of planetary systems only one planet has environments suitable for the process of biological evolution. Then many millions of planets on which life is possible should exist within the confines of our Galaxy.

Astronomers tell us that our Galaxy is merely one of millions upon millions of other galaxies.

From what we know about the Universe it can be asserted that living things must have a feature common to all of them, namely, that any living being is a tremendously complicated aggregate of chemical compounds of hydrogen, oxygen, carbon, nitrogen, phosphorus and many other elements which are the same for the whole universe.

Another thing should also be admitted; in case life exists on other planets, it may have quite unexpected forms, absolutely uncomparable with those on our Earth.

Still another point not to be ignored is that the level of civilization of even those intelligent beings on other planets who in their development stand closest to man will drastically differ from our own.

The existence of man on the Earth is only an instant if measured on the infinite scale of cosmic time. Therefore on some planets intelligent life may be well ahead of the level we have reached.

Possibly, some technically advanced civilizations are already sending signals into outer space to inform other worlds about their existence and establish communication with them. It is not improbable that there exist such civilizations for which communications with other planets have long become a sort of conventional "cultural exchange".

Still, it would be a mistake to consider that to contact inhabitants of other worlds is an easy task. On the contrary, this task is quite difficult, since the theory of relativity imposes a limitation on the speed with which information can be transmitted. Even such a tremendous speed as that of light, i.e. 300 000 km/s, is not a very large value on the cosmic scale.

Attempts have already been made to receive signals from intelligent beings, inhabitants of the Universe. To this end, a "search" of the Universe was tried out on the frequency of 1420 MHz corresponding to the 21-cm wavelength. The choice of this particular wavelength was dictated by the following reasons.

The idea came from the American scientists G. Cocconi and Ph. Morrison who in the 1940s theoretically predicted that neutral atoms of hydrogen found under the conditions of interstellar space must emit the 21-cm spectral line. This wavelength is associated with the passage of the hydrogen atom from one energy state in which the magnetic moments of the hydrogen nucleus and electron are parallel into another, in which they are antiparallel.

Thus, the frequency standard was prompted, as it were, by the Nature itself. Scientists suggested that artificial signals in the Universe should be sought for at the frequency corresponding to the 21-cm wavelength. There can be no doubt that any technologically advanced civilization must have discovered this frequency in the spectrum of cosmic radio radiation as well.

In May—July, 1960, a program for the search of signals of extraterrestrial civilizations (Project Ozma) was carried out for 150 hours, at the American Radio Astronomy Observatory at Green Bank in West Virginia under the leadership of the radio astronomer F. Drake. The search was conducted with specially developed equipment focused on the nearest stars Tau Ceti and Epsilon Eridani which are believed to have planetary systems and are relatively not very remote from the Sun, the distance being about 11 light years. Though the observations by a 27-metre radiotelescope were very careful, no signals of an artificial character were received. It should, certainly, be taken into consideration that chances of success were not great. Yet it was decided to continue the research.

A similar experiment was conducted in the Soviet Union at Sternberg Astronomical Institute. At present the search of signals of extraterrestrial civilizations is under way at Gorky State University Research Institute of Radiophysics. No positive results have been obtained so far. But one must have patience when searching for "intelligent" signals in outer space and, perhaps, it will take centuries from now.

But are we going the right way? Maybe we exaggerate the importance of the 21-cm wave in space communications and are led astray by the theoretical predictions that precisely this wave is a "space bridge" to extraterrestrial civilizations?

Soviet scientists checked the initial premises and came to the opposite conclusions. The wave emitted by neutral atoms of hydrogen is least likely to carry signals of extraterrestrial civilizations! This is just the frequency band where maximum cosmic noises should be expected: all the interstellar hydrogen is humming on this wave! A signal sent by any intelligent civilization will be inevitably absorbed by this noise. One should seek for such a band where the noises would be minimum, while the conditions for the transmission and reception of signals would be most favourable. This problem is discussed by the Soviet astronomer N. Kardashev in his paper devoted to communication with extraterrestrial civilizations and published in Astronomical Journal (41, Issue 2, 1964).

If N. Kardashev is right in his conclusions, an attempt at communicating with extraterrestrial civilizations on the 21-cm wave will probably bring a very disappointing result.

But radio communication is not the only means of contacting the inhabitants of other worlds.

Many scientists are of opinion that it will be most reasonable and promising to use lasers for establishing communication with extraterrestrial civilizations. Such an application of lasers may open an era of interstellar communication.

The American scientists Townes and Schwarz were the first to suggest the use of lasers for space communications in an article on this subject published in 1961. In this article they discuss the possibility of detecting signals sent by means of a laser beam from a planet of a star which is at a distance of tens of light years from us. It is assumed that messages are transmitted by a society which has reached approximately the same level of development as we have. Such signals sent to us can well be detected by our present-day telescopes and spectrographs.

In their discussion of the problem, the authors consider two systems of lasers, *system a* and *system b*, as the main equipment. The characteristics of *system a* are as follows. Its power is 10 kW, it works CW with the wavelength of about 5000 Å, the frequency band width is of the order of 1 MHz, and the diameter of the reflector is 500 cm (the diameter of the biggest now available reflecting telescopes). The beam width (the ratio of the wavelength to the reflector diameter) is 10^{-7} radian.

System b consists of 25 lasers with the characteristics of *system a*. The effective aperture is 10 cm, the beam width is $5 \cdot 10^{-6}$ radian. All the lasers of this system are oriented strictly in one direction with an accuracy of up to the beam width.

System a must be mounted beyond the Earth's atmosphere, say, on the Moon or on an artificial satellite, since otherwise the effective width of the beam will be considerably reduced by the atmospheric turbulence.

As to *system b*, it must be capable of effectively operating when mounted on the surface of a planet having an atmosphere similar to that of our Earth.

The beam sent to us from a neighbouring star can be detected if its intensity is sufficient and this beam differs from the background, that is from the light radiated by the star.

The laser beam emitted by *system a* may be seen with a naked eye from the distance of 0.1 light year, and through binoculars, from the distance of 0.4 light year. From the distance of 10 light years the radiation emitted by *system a* may be detected visually with a 500-cm telescope, or photographed by using a conventional method with an exposure time of 1 minute.

For *system b* the exposure time is considerably longer. Over the distances specified for *system a*, it will be difficult to detect the signal with a naked eye or a telescope. Some scientists are of opinion that, contrary to the assertions of Townes and Schwarz, this *system b* is altogether unfit for interstellar communications.

Is there a possibility of discriminating between the emission of a laser and that of a star? One of the distinctive features of a laser emission may be its monochromaticity. For example, if the distribution of emission of a star is the same as that of our Sun, then according to calculations, in the vicinity of the laser operation wavelength of 5000 Å the spectral density of the laser emission for the band of 2 MHz would be 25 times that of the star. Should the laser operate in the ultra-violet or infra-red region of the spectrum, the laser emission intensity would still further exceed that of the star, since the emission of the latter (assuming it to be similar to the solar one) is particularly pronounced within 5000 Å.

The possibility of detecting a laser beam against the background of the star emission will be increased by hundreds and thousands of times if the laser is located beyond the atmosphere. In such a case observations should evidently be conducted by using narrow-band filters.

Another method for making the laser beam detectable against the star emission background is to modulate this beam, and this will evidently take place during the transmission of messages. A modulated laser beam can be detected, by using correlation methods, even against background noises whose level is commensurable in magnitude with the laser emission.

In view of all this lasers can be regarded as quite suitable means for establishing interstellar communication. The idea of communicating with extraterrestrial civilizations appears still more realistic when one takes into account that before long much more powerful lasers will be devised.

The achievements made in many branches of science and technology (such as astronomy, biology, cybernetics, information theory, radiophysics and radioengineering, development of space) are so fundamental that now the problem of communications with extraterrestrial civilizations is the subject of top-level scientific discussions.

In 1970 the USSR Academy of Sciences and the National Academy of Sciences of the USA reached an agreement that a joint conference should be held on this problem. The Conference took place in September 1971 at Burakan (the Armenian Soviet Socialist Republic), and its participants were scientists not only from the Soviet Union and the United States, but from other countries as well. This was the second meeting of scientists in the Soviet Union, devoted to the discussion of this important problem. The first All-Union Conference was called as early as 1964 expressly to consider the problem of extraterrestrial civilizations and possibility of establishing contacts with them. This Conference was held also at the Burakan Astrophysical Observatory of the Academy of Sciences of the Armenian Soviet Socialist Republic.

Astronomers, physicists, biologists, antropologists, historians, sociologists, philosophers, linguists, experts in information and communication theory—such is the list of those who took part in the work of the International Conference on the Problem of Communication with Extraterrestrial Civilizations in 1971. Among its participants there were such prominent Soviet scientists as Academicians V. Ambartsumian and V. Ginzburg, Corresponding Members of the USSR Academy of Sciences V. Siforov, V. Troitsky and I. Shklovsky. American science was represented, in particular, by the well-known physicists Nobel Laureate Professor Charles Townes, Doctor Freeman Dyson, astronomer Francis Drake, expert in cybernetics Professor Martin Minsky, biologist and astrophysicist Doctor Carl Sagan. Great Britain was represented by Nobel Laureate Professor Francis Crick. The aim of the Conference was to estimate the actual state of the problem and delineate the ways of further activities.

Opening the Conference, Academician V. Ambartsumian said: "Certainly, an opinion may be voiced that the discussion about extraterrestrial civilizations and communication with them is premature, since so far there is no direct concrete evidence of the existence of extraterrestrial civilizations. But the initiators of the Conference consider that an active search for such evidence is necessary, as well as a comprehensive theoretical investigation of the problem, based on all the data of modern astronomy, planetology, biology and sociology".

The participants of this round-table Conference discussed many interesting questions, both of problematical and cognitive character: modern views on the origin of planets and methods of their detection at the nearest stars; the possibility of the existence of life on cool planets, on planets without stars, and even in the interstellar medium (complex organic compounds having been discovered by radioastronomic methods in the interstellar medium); the possibility of life transport from one planet to another; the influence of submarine volcanoes on the origin of life on the Earth; the role of chance in the process of the origin of life. An interesting discussion was run around the problem of the origin of reason. It was pointed out that complicated social life is a prerequisite for the origin of reason. However, a combination of numerous factors is required for the development of a thinking being. Ideas were expressed on the existence of artificial reason within the confines of the Universe.

G. Marks (Hungary) made a report on interstellar flights. He considered the motion of an interstellar spaceship under the effect of light pressure of a laser beam emitted from a home-planet. The main difficulty with such a method of communication is that the spaceship can be sent back only by other civilization. "Insofar as we are not yet in possession of such machinery", concluded the speaker in half a joke causing general animation of the audience, "nobody has flown to us".

In the discussion of the problem of searching for information signals F. Drake emphasized that electromagnetic waves are the only rapid, effective and economically expedient means to establish communication with extraterrestrial civilizations. In spite of considerable difficulties, a sustained systematic search for signals from other worlds, carried out with an extensive use of modern computers for the analysis of the incoming radiation, in his opinion, can be successful.

N. S. Kardashev posed the question of seeking for such methods of transmitting signals, which would require minimum consumption of energy per pulse. He said that in one case, when we are in search for a sender whose location in space is absolutely unknown, the frequency range of 10^9 to 10^{11} Hz may prove optimal, while in another case infra-red and submillimetre regions may be best.

Ch. Townes presented arguments in favour of the use of lasers. In his opinion, lasers may well compete with radioengineering communication means over distances less than 5000 light years and are especially promising in the search for signals from those nearest stars which are at a distance of a few light years or several tens of light years from the Earth. However, he was against uncompromising preference to any one method for establishing contacts with extraterrestrial civilizations and said that various technological possibilities should be exploited and, particularly, different frequencies should be tried.

Ph. Morrison pointed out that the quality of the information received and not its amount would be of primary importance to us. The first call signal would be the most important step.

B. Oliver (USA) emphasized that the problem of contacting extraterrestrial civilizations was a matter of great concern; in this connection he informed the Conference about a program for the search of signals of extraterrestrial civilizations, being developed in the United States and named "Cyclops Project". The Cyclops Project is based on the use of a multielement aerial system having up to 10 000 mirrors of 20 to 30 metres in diameter with a complicated communication system. B. Oliver stressed that such a multielement system was much easier to realise than one large aerial of an equivalent surface area. Besides, the potential of such system can be built up gradually, this factor being of particular importance. For frequency search of signals the project envisages the use of receivers with up to one million of frequency channels.

Yet, to establish communication with intelligent inhabitants of a planet is only half of the business, since for getting in touch with them, for talking with them, a language must be elaborated. Probably, drawings would be most expressive in information exchange, so that initial messages might be graphical. Certain notions derived from common physical laws could be communicated. For instance, it would be possible to transmit prime numbers which remain primes in any scale of notation. Another example of what is common among inhabitants of different worlds is the structure of atoms.

An example of what a message received from the depth of the Universe might be is given by B. M. Oliver in his article "Interstellar Communication" ("Interstellar Communication". A Collection of Reprints and Original Contributions. A.G.W. Cameron, Editor. W. A. Benjamin, Inc. New York, Amsterdam, 1963, pp. 302-305). He writes: "Let us assume that after years of futile listening we receive a peculiar series of pulses and spaces from ε Eridani. The message is repeated every 22 hours and 53 minutes, apparently

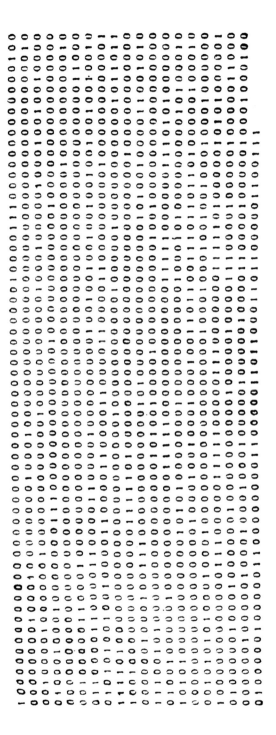

Fig. 92. Imitation message from outer space

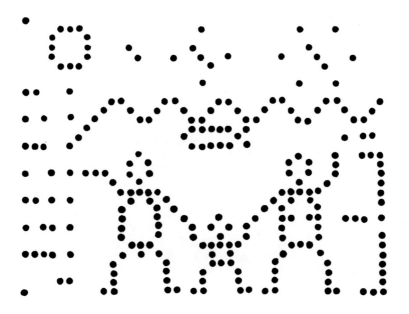

Fig. 93. Same message rearranged

the length of their day. The pulses occur at separations which are integral multiples of a minimum separation. Writing ones for the pulses and filling in the blanks with the appropriate numbers of zeros we get the binary series shown in Figure 3 (our Figure 92). It consists of 1271 ones and zeros. 1271 is the product of two primes 31 and 41. This strongly suggests that we arrange the message in a 31×41 array. When we do so, leaving blanks for the zeros and putting down a dot for each pulse we get the non-random pattern of Figure 4 (our Figure 93).

Apparently we are in touch with a race of erect bipeds who reproduce sexually. There is even a suggestion that they might be mammals. The crude circle and a column of dots at the left suggests their sun and planetary system. The figure is pointing to the fourth planet, evidently their home. The planets are numbered down the left-hand edge in a binary code which increases in place value from left to right and starts with a decimal (or rather a binary) point to mark the beginning. The wavy line commencing at the third planet indicates that it is covered with water and the fish-like form shows there is marine life there. The bipeds know this, so they must have space travel. The diagrams at the top will be recognized as hydrogen, carbon, and oxygen atoms, so their life is based on a carbohydrate chemistry. The binary number six above the raised arm of the right figure suggests six fingers and implies a base twelve-number system".

If we succeed in getting in touch with some extraterrestrial civilization, we shall be able to receive much valuable information from it. Suppose this civilization is 10 000 years ahead of us in its development. Then the technical level it has reached must be very high indeed. Just imagine, what a powerful source of technical progress for us such information might be! After the establishing of communication personal contacts will probably be needed.

The next stage in communication with extraterrestrial civilizations will be space flights.

SPACESHIP OF THE FUTURE

Our age is called the Space Age. Now that man has mastered near-terrestrial space, his will incessantly urges him further, towards the planets of the Solar System, towards other distant worlds.

Space journeys throughout the Solar System and beyond it can be achieved only through the development of spaceships moving at a speed close to that of light.

At present there exists a large variety of projects of future spaceships, as well as of atomic, photon, plasma, quantum and ion engines for them. The use of rocket engines utilizing the emission of alpha particles has also been suggested. A very interesting project of a solar sail rocket engine is based on the phenomenon of solar pressure which was predicted by Maxwell and later proved to exist by Lebedev. In this case the role of "cosmic wind" to "fill the sails" of the spaceship will be played by solar rays.

Obviously we have a long way to go before all these projects can be realised. The very idea on which many of them are based is somewhere between the real and the fantastic, between the possible and the impossible. Yet, there are no limits to what man can achieve, so that the boundary between the reality and fancy is merely provisional. The reality of to-day boldly trespasses the frontiers of yesterday's fancy. The impossible becomes the possible.

The appearance of lasers capable of providing a high-power directed beam suggested the idea of a spaceship propelled by a laser engine. Indeed, if the idea of using solar pressure for imparting acceleration to an object free in space is right, then a laser which is a powerful source of radiation can procure reactive force for imparting the necessary acceleration to our spaceship. In any case, such an engine is not less realistic than the engines contemplated in all those projects of which we spoke before.

In principle, a laser engine does not differ from any other reactive engine. Its reactive thrust is created by photons, i.e. by light quanta which, as we know, possess a mass.

If we take into account that the engine of the spaceship will be accelerated steadily and for a long period of time, its speed can be ultimately brought close to that of light.

LOOKING AHEAD

Our description of laser applications in various branches of science and engineering is drawing to an end. But, naturally, we could not make the picture complete. A lot of interesting, perhaps unexpected and still more valuable applications are in store for these devices. In view of the headlong advance of technology nowadays, it would be quite difficult to predict possible future uses of lasers. We shall mention only some of them, which even to-day seem interesting and sound promising.

Scientists are carrying extensive research, trying to harness the thermonuclear reaction. The day the scientists learn to control this reaction the mankind will be in possession of inexhaustible power resources. The thermonuclear reaction requires a very high temperature amounting to tens of millions of degrees. So far such a temperature can be attained through an explosion of a hydrogen bomb, this being the only way of ensuring it artificially. With lasers coming of age, an idea was put forward that in the focus of a laser beam a plasma bunch could be heated to temperatures sufficient for the thermonuclear synthesis. By focusing a giant pulse of a laser on a solid target, it is possible to obtain plasma having an extremely high temperature.

Not long ago a series of unique experiments were completed at the Lebedev Physics Institute of the USSR Academy of Sciences. The team of researchers headed by A. Prokhorov in one of the experiments conducted with a ruby laser succeeded in obtaining dense plasma having a temperature of about 500 000°K.

Another experiment demonstrated that in case of laser pulses of 30 MW more than 10^{14} ions had energy from 1.0 to 10 keV. The results of measurements showed the temperature of plasma in the focus of a 30 MW laser beam to be about 1 000 000°K.

These experiments give grounds to expect that the laser will be the very match we need for kindling the controlled thermonuclear reaction.

An interesting suggestion was made to use laser beams for correcting the trajectories of artificial satellites. The pressure of light has been known to exist since the outstanding Russian scientist Lebedev proved it experimentally in 1900. This effect is resorted to when a high-output laser beam is directed from the Earth to the satellite for correcting its trajectory. The light pressure exerted by the beam on the satellite will urge the latter upwards and compensate for those inevitable losses in the altitude of the satellite after each its revolution, which are caused by the resistance of cosmic particles, however small this resistance may be. So the satellite's lifetime in orbit can be substantially prolonged. The same effect can be employed when creating manned orbiting space stations.

Superpowerful light beams might safeguard the spaceship during its interstellar flight against collision with individual meteors. Though these meteors are quite minute solid particles, they move at tremendous speeds and the impact force they can develop comes to scores of tons—a thing not to be trifled with! A powerful laser beam could make the meteor particle to swerve from the path of the spaceship and prevent the collision dangerous for the astronauts.

If the spaceship happens to encounter a meteor stream, the consequences will be disastrous: the ship will perish. A laser beam serving as a radar can timely detect such dangerous areas of meteor streams and help in choosing most safe route for the flight.

A spaceship laser radar will help to determine the distance to celestial bodies, as well as to improve the trajectory and route of the flight. Small dimensions and a relatively low input power required by the laser will contribute to its applicability as spacecraft equipment. As far back as the thirteenth century Roger Bacon put forward an idea that energy could be transmitted with the help of a light beam. He suggested a system of mirrors which "would be worth a whole army against the Tartars and Saracens". Now this idea of transmitting energy by means of a light beam can be realised with the aid of lasers. It will probably be used for transmitting energy to those places which are difficult of access and where energy cannot be transmitted by conventional methods.

Recently a project of an electron accelerator employing a laser was suggested. The idea is as follows. A cylindrical pipe made from a material used in lasers as their active medium is excited by pumping radiation through an interference filter placed on its external surface. Oscillations are generated inside the pipe. According to calculations, optical energy reaches a maximum of 10 kW/cm^2. Electrons are accelerated by a powerful electric field (about 10^9 V/m).

Evidently, lasers will be widely employed in rockwork and mineral mining. These devices will also come to the aid of ice-breakers: a laser beam will crush the ice and clear the way for the ship.

In the nearest future lasers may find extensive application in building engineering. With their help separate blocks or bricks can be fused together into a strong monolithic wall. Many materials which are now considered infusible will be welded together by a laser beam. With lasers, buildings will be erected much faster and their quality will be far better.

Soon a laser beam will become an artist's tool: it will be used for engraving patterns on decorative ceramics and even for making sculptures.

If we give reign to our imagination, we shall see new vistas opened by lasers for three-dimensional colour cinematography and television. The stereoscopic effect which can be achieved with lasers will bring about radical changes in cinema and television technique. Dynamic three-dimensional colour images demonstrated on the screens of motion-picture or television theatres will be not inferior to natural vision. Laser communication lines which will be similar to the now available radio relay or coaxial transmission lines, will interlink all the towns on the continents. Tremendous potentialities of such information transmission channels, capable of coping equally well with any kind of information, will allow outside television broadcasts of theatre shows, sporting events, festivals or meetings with astronauts to any point of the globe, and the scenes will lose nothing as regards their natural plasticity, colour or dimensions. Casting our mind's eye on the towns of the future, we see breathtaking pictures which their inhabitants will be able to enjoy in the evenings: the sky above the town, when covered with clouds, haze, or, perhaps, with some artificially created semitransparent medium, will be converted into a colossal screen, on which magnificent three-dimensional "laserovision" shows will be presented, visible to any town-dweller wherever he may be. This will be like a mirage, only much more grand than anything we can see in nature, since man will be its creator. The scenes above the towns will carry the audience to sea shores, tropics or high latitudes, or even to other worlds in deep space.

In conclusion we should like to mention one more idea about the prospective use of lasers, which was put forward by I. Shklovsky in his book "The Universe, Life and Reason" (2nd Edition, "Nauka" Publishers, Moscow, 1965). In their stellar work astronomers often observe outbursts of stars which are called "supernovae". These outbursts resemble giant explosions of cosmic bodies. Assuming the existence of highly developed extraterrestrial civilizations in far-out space, these giant explosions may be supposed to be of artificial character: highly developed intelligent beings may explode neighbouring stars to replenish the stock of heavy elements they need. But how can this be brought about? How can one explode such a colossal heavenly body as a star? It turns out that a star can be exploded with the help of... a laser! Suppose that intelligent beings are in possession of superpowerful lasers operating in the range of gamma-radiation with a wavelength of, say, 10^{-10} cm. If the laser aperture is 10 m, then the beam divergence angle will be only two hundred millionths of an angular second. If the star to be exploded by such a unique procedure is at a distance of 10 light years, the diameter of the "spot" from the "gamma-laser" beam as it reaches the surface of this star will not exceed 10 km.

The flux of gamma-radiation required to fall on the surface of the star for initiating its nuclear explosion should be about 10^{10} erg/(cm$^2\cdot$s). Such a radiation flux can be produced by a laser system with an output power of about 10^{12} kW. This figure is 1000 times higher than the total amount of power which can be produced by all the sources of energy available to our modern civilization. Yet, taking into account the headlong advances we are making, it would not be unreasonable to consider such energies to be quite available to highly developed extraterrestrial civilizations.

This example once again demonstrates unlimited potentialities of lasers. With the discovery of lasers mankind has made a big step forward in its technical development. It will make still greater strides by using these devices in various ways.

Printed in the United States
21710LVS00004B/106